T0073195

The Finite Quantum Many-Body Problem

Selected Papers of Aage Bohr

World Scientific Series in 20th Century Physics

Published

Vol. 45 *The Finite Quantum Many-Body Problem: Selected Papers of Aage Bohr*
edited by Ricardo A. Broglia

Vol. 44 *Symmetries in Nature — The Scientific Heritage of Louis Michel*
edited by T. Damour, I. Todorov and B. Zhilinksii

Vol. 43 *Nambu: A Foreteller of Modern Physics*
edited by T. Eguchi and M. Y. Han

Vol. 42 *Selected Papers of K. C. Chou*
edited by Y.-L. Wu

Vol. 41 *Many-Body Theory of Molecules, Clusters, and Condensed Phases*
edited by N. H. March and G. G. N. Angilella

Vol. 40 *Murray Gell-Mann — Selected Papers*
edited by H. Fritzsch

Vol. 39 *Searching for the Superworld — A Volume in Honour of Antonino Zichichi on the Occasion of the Sixth Centenary Celebrations of the University of Turin, Italy*
edited by S. Ferrara and R. M. Mössbauer

Vol. 38 *Matter Particled — Patterns, Structure and Dynamics — Selected Research Papers of Yuval Ne'eman*
edited by R. Ruffini and Y. Verbin

Vol. 37 *Adventures in Theoretical Physics — Selected Papers with Commentaries*
by Stephen L. Adler

Vol. 36 *Selected Papers (1945–1980) with Commentary*
by Chen Ning Yang

Vol. 35 *A Career in Theoretical Physics (Second Edition)*
by P. W. Anderson

Vol. 34 *Quantum Mechanics in Phase Space: An Overview with Selected Papers*
edited by C. K. Zachos, D. B. Fairlie and T. L. Curtright

Vol. 33 *Formation and Evolution of Black Holes in the Galaxy — Selected Papers with Commentary*
edited by H. A. Bethe, G. E. Brown and C.-H. Lee

Vol. 32 *In Conclusion — A Collection of Summary Talks in High Energy Physics*
edited by J. D. Bjorken

Vol. 31 *From the Preshower to the New Technologies for Supercolliders — In Honour of Antonino Zichichi*
edited by B. H. Wiik, A. Wagner and H. Wenninger

Vol. 30 *Selected Papers of J. Robert Schrieffer — In Celebration of His 70th Birthday*
edited by N. E. Bonesteel and L. P. Gor'kov

Vol. 29 *Selected Works of Emil Wolf (with Commentary)*
edited by E. Wolf

Vol. 28 *The Legacy of Léon Van Hove*
edited by A. Giovannini

Vol. 27 *Selected Papers of Richard Feynman (with Commentary)*
edited by L. M. Brown

For information on Vols. 1–26, please visit http://www.worldscientific.com/series/wsscp

World Scientific Series in 20th Century Physics Vol. 45

The Finite Quantum Many-Body Problem

Selected Papers of Aage Bohr

editor

Ricardo A Broglia

Niels Bohr Institute, Denmark

World Scientific

NEW JERSEY • LONDON • SINGAPORE • BEIJING • SHANGHAI • HONG KONG • TAIPEI • CHENNAI • TOKYO

Published by

World Scientific Publishing Co. Pte. Ltd.
5 Toh Tuck Link, Singapore 596224
USA office: 27 Warren Street, Suite 401-402, Hackensack, NJ 07601
UK office: 57 Shelton Street, Covent Garden, London WC2H 9HE

Library of Congress Control Number: 2019030183

British Library Cataloguing-in-Publication Data
A catalogue record for this book is available from the British Library.

World Scientific Series in 20th Century Physics — Vol. 45
THE FINITE QUANTUM MANY-BODY PROBLEM
Selected Papers of Aage Bohr

ISBN 978-981-120-813-3

For any available supplementary material, please visit
https://www.worldscientific.com/worldscibooks/10.1142/11495#t=suppl

Dedicated to the memory of
Pier Francesco Bortignon,
collaborator, friend, reference point
in the scientific endeavor.

Foreword

Aage Bohr (1922–2009), the only of Niels Bohr's sons who also became a physicist, grew up surrounded by some of the most important figures in twentieth century science, among them Hendrik Kramers, Oskar Klein, George de Hevesy, Werner Heisenberg, and Wolfgang Pauli. In 1943, at age 21, he fled from Nazi-occupied Denmark together with his father and became his right-hand man. He returned in 1945, studied physics at the University of Copenhagen, worked at Columbia University in 1949–1950, returned to Copenhagen again—to the University's Institute for Theoretical Physics his father had founded in the early 1920s—and became a Professor there in 1957. In 1963, shortly after his father's passing, he became the director of what soon (in 1965) would be named The Niels Bohr Institute. His work and that of his collaborators, which centered on nuclear *structure* physics, gave rise to what Gerald E. Brown would later dub the "Copenhagen school of nuclear physics,"[1] putting Aage's contributions to science in line with those of the man behind the Copenhagen school of atomic and quantum physics of the 1920s—his father.

This volume assembles selected writings of Aage Bohr—some well known to practitioners in the field, others less so—and makes them available to a broader audience, not only of physicists, but also of historians and philosophers of science and other interested readers. In collecting and selecting these writings and in complementing them with a substantial and highly accessible introduction that, besides explaining the key physical concepts at play, also does not eschew their historical and philosophical dimensions, Ricardo A. Broglia—himself a close co-worker of Aage's over several decades—has achieved a synthetic view on an important strain of Aage Bohr's research: the finite quantum many-body problem of the nucleus. While in principle always there for everyone to see, bringing together crucial pieces of the mosaic scattered in Aage Bohr's many writings, Broglia reveals the conceptual and eminently physical motivations underlying Aage Bohr's research, and invites the readers of this volume to look and see for themselves.

As director of the Niels Bohr Archive—an institution devoted to preserving the heritage not only of Niels Bohr's but also of Aage Bohr's and their co-workers' papers—my own interest in Aage Bohr's physics is first and foremost historical. The Niels Bohr Archive is currently preparing for a large-scale historical research project, which will spell out the developments in nuclear structure physics in Copenhagen and thus contextualize the work done

[1]G. E. Brown, "Fly with Eagles." *Annual Review of Nuclear and Particle Science* 51, no. 1 (2001): 1–22, on 12.

by Aage Bohr and his collaborators within the broader history of physics in the twentieth century. The present volume—although not officially part of it—will serve as an important stepping stone for this project.

Christian Joas
Niels Bohr Archive
Copenhagen, November 2018

Contents

Overview

The papers of Aage Bohr reprinted in the present volume were selected because they provide a clear and beautiful synthesis of his epoch-making contribution to our present-day understanding of nuclear structure in a technically unburdened, deeply physical fashion. In a very real sense they can not only be used to understand nuclear structure, but also as a source of precious insight to further develop the subject, in particular concerning its connection with the broad field of many-body physics.

The publication date of the scientific papers span a time interval of two decades. Starting with the first reference on the interplay between single-particle and collective motion [I], and ending with the Nobel lecture [VIII] and the related Varenna lectures [IX] recounting the extension of the unified model to encompass also gauge space and pairing modes. Paper [X] provides a glimpse into the network of scientific collaborations and human relations that lie at the basis of this achievement.

The list of papers selected, although not meant to be univocal is, nonetheless, rather natural. The red thread going through them all is that of providing the physical concepts and the grand view lying at the basis of Aage Bohr's development of the unified collective and individual-particle description of the atomic nucleus, where pairing dynamical modes are treated on par with surface and density vibrations, and in which reaction processes, in particular those involving the transfer of one- and two-nucleons, are the specific probes.[1] The consequences of the physics contained in the papers selected are still playing an important role in the ongoing cutting-edge research lying at the forefront of nuclear structure studies, let alone in the development of a unified field theory of structure and reactions.

Before proceeding with the discussion of the main topics, namely spontaneous broken symmetry, elementary modes of excitation and nuclear field theory (NFT), let us give a brief overview of features relating to the selected papers. Quoting from [I]:

> The individual particle motion ... has accounted successfully for a large number of nuclear properties ... However, nuclear matter appears to have some of the properties of coherent matter which make it capable of types

[1] Concerning the contributions made to these developments by Ben Mottelson and other collaborators, see the references and the explicit notes to be found in the different papers. The role played by the collaboration with Ben Mottelson in the scientific production of Aage Bohr is quite large, as testified by the many papers they coauthored and by their monograph *Nuclear Structure, Vols. I and II*, Benjamin (1969, 1975). The papers selected in the present compilation reflect, nonetheless, some aspects of the unique approach and ideas Aage Bohr brought and contributed to what can be called the second discovery of the atomic nucleus, a laboratory to explore quantum mechanics within the scenario of finite many-body systems.

> of motion for which the effective mass is large as compared with the mass of the single nucleon

and then from [II]:

> The motion of the individual particles is strongly affected by the collective oscillations, and the nuclear properties are determined by the interplay of these two basic types of motion.

One thus finds in [I, II] the elements at the basis of the particle-vibration coupling and its eventual evolution into a field theoretical description of nuclear structure.

To elaborate on the above quotations, let us use Schwinger's prologue to his Quantum Mechanics Lectures written many years later "... what we call physically an electron is only partially to be associated with the electron field alone. It is also partially to be associated with the photon field, because the two are in interaction ... this program of renormalisation, to mention the technical term, as it was applied specifically to the case of electrons and photons, through the development of what is called quantum electrodynamics, led to the description ... of the finer features of electron and photon behaviour..." [1].

Concerning the content of article [III], it extends the role, the interweaving between the different modes of excitation play in the nuclear spectrum, to deal with anharmonicities of phonon states resulting from the coupling of one- to two-phonon excitations. This gives rise to what eventually became known as the doorway mechanism in the damping of the giant dipole resonance.[2] Translated into condensed matter language, it corresponds to the inhomogeneous damping of the plasmon. Furthermore, in this paper the transfer quantum number β is introduced which together with multipolarity, parity, spin and isospin, characterizes the nuclear response. Concerning β, we read:

> For the particle excitons, we have $\beta = \pm 1$, for a particle or hole. For the bosons, $\beta = 0$ for a mode associated with oscillations in the average field, or a particle-hole mode, while $\beta = +2(-2)$ for an exciton which involves the addition (or subtraction) of two nucleons to the nucleus. The "excitation" which takes us from the ground state of an even nucleus to the ground state of the next even nucleus is of this type (pairing mode with $\lambda\pi = 0^+$). Other pairing modes have so far received only little attention.

In the last sentence, Aage Bohr refers to isovector and to multipole pairing vibrations (with angular momentum λ and parity π, $\lambda^\pi = 2^+, 4^+, 6^+ ...$), the latter constituting a subject which is today an active area of research. Specifically regarding the calculation of the corresponding absolute two-particle transfer differential cross-sections, quantities which are needed to quantitatively analyze double charge exchange processes as an analogue, as far as the nuclear matrix element is concerned, to neutrinoless beta decay[3] (see e.g. [2]).

[2] The concept of doorway state was introduced in the work of Block, B. and Feshbach, H. (1963), Ann. Phys. 23:47. The general formulation of the subject was worked out in Feshbach, H., Kerman, A., and Lemmer, R. H. (1967), Ann. Phys. 41:230.

[3] Within this context see the recently published Proceedings of the Conference on Neutrino and Nuclear Physics (CNNP2017) 17–21 October 2017, Catania, Italy, Journal of Physics: Conference Series Volume 1056 (2018).

The transfer quantum number β connects directly with the variety of specific probes exciting $\beta = 0$ (inelastic scattering and Coulomb excitation), $\beta = \pm 1$ (one-particle stripping e.g. (d, p) and pickup, e.g. (p, d) reactions) and $\beta = \pm 2$ (two-nucleon transfer processes, e.g. (t, p) and (p, t) reactions) nuclear modes.

Paper [IV] gives a profound and virtually comprehensive account of the elements needed to work out a field theory with an overcomplete basis of states, as in the nuclear case, in which phonons are made out of the same degrees of freedom as those associated with single-particle motion (see Sec. 5 on spurious degrees of freedom). Figure 6 of this article provides a remarkable synthesis of structure (diagrams (e) and (f)) and reactions (diagrams (a)–(d)) Feynman diagrams, to be eventually found at the basis of nuclear field theory. Furthermore, a variety of facets associated with pairing modes are discussed in Sec. 6.

Paper [V] introduces, aside from the concept of ground state correlations associated with pairing vibrations, NFT Feynman diagrams describing the coupling of single-particle motion and pairing vibrations (Fig. 6 of the paper).

Paper [VI] deals with the concept of nuclear rotations in general and in particular with the specific probe of pairing rotational modes: two-nucleon transfer reactions. A parallel with Cooper pair tunnelling between weakly coupled superconductors is made (Josephson effect).

In paper [VII], the variety of subjects which, through their interweaving generate the red thread going through the articles selected are all present, namely quantum theory of the many-body problem, nuclear field theory and collective modes involving nuclear transfer. The corresponding description, which does not make use of a single formula, becomes progressively more profound and detailed. It also reflects the results which collaborators dealing with theoretical and experimental aspects of the subject had obtained in the meantime ([3–7] and refs. therein).

Papers [VIII] and [IX] concentrate on rotational motion which also encompasses pairing rotational bands, and the interplay of rotations in 3D space and of pairing correlations leading to the superfluid-normal phase transition.[4]

The main results from the topics of static and dynamic pairing dealt with in the papers selected, have resulted in the test of the corresponding many-body theory in terms of individual states, as Aage Bohr predicted. This feat, not possible in condensed matter, has shed new light onto BCS and associated Cooper pair Josephson tunneling. Only recently has it become possible to routinely calculate absolute differential cross-sections within 10% error, an accomplishment similar to that of being able to accurately calculate the maximum value of Josephson's current.

Concerning NFT, the insight into nuclear structure provided by the selected papers has had an important impact. That of helping to understand how, through renormalization of single-particle and collective modes, and of both energies and radial wave functions, an eventual exact solution of the nuclear many-body problem may be worked out in terms of Feynman diagrams having as single input the nucleon–nucleon interaction (for a partial summary see [8, 9] and refs. therein). Within this context, the heritage of [I–IX] is today helping to attack and solve central open issues of the nuclear many-body problem.

---

[4]Phenomena connected to the Meissner effect and to the superconducting–normal phase transition triggered by a critical magnetic field in low temperature metallic superconductors.

A glimpse into how such achievements could be accomplished is provided by the speech Aage Bohr gave on the occasion of the 50th anniversary of the inauguration of Universitets Institut for Teoretisk Fysik (UITF, University's Institute for Theoretical Physics), known today as The Niels Bohr Institute [X]. A grand project, a lively cooperation — deeply scientific and strongly human — with both local and international practitioners and collaborators, and a strong interplay between theory and experiment. An equally important accomplishment of Aage Bohr was to continue to secure a place of prominence for UITF among the world's centers of excellence in physics research for two decades after his father's death in 1962, when he took over the direction of the Institute.

Reading the papers selected, in which Aage Bohr conveys his understanding of the nuclear structure, not only to his nuclear colleagues, but also to many-body practitioners in a physically complete and technically unburdened fashion, one achieves a clearer and deeper understanding of his ideas. And in the process, apprehend nuclear structure within the broad framework of quantum mechanics of systems with many degrees of freedom, closer to its emergence.

1. Incipit

In an attempt to set up the stage, I quote the assessment by Philip W. Anderson [10] concerning work contained in the papers of Aage Bohr collected in the present volume: ... "It is fascinating ... that nuclear physicists stopped thinking of the nucleus as a featureless, symmetric little ball and realised that ... it can become [American] football-shaped or plate-shaped. This has observable consequences in the reactions and excitation spectra that are studied in nuclear physics ... this ... research... is as fundamental in nature as many things one might so label".

And again, many years later [11], describing aspects of activity of theoretical physicists in the 1960s and early 1970s: "... the story of broken symmetry[5] ... is a heartening story of one of those rare periods when the fragmentation of theoretical physics into condensed matter, nuclear and particle branches was temporarily healed and we were all consciously working together in exploring the many quantum consequences of the idea of broken symmetry".

2. Spontaneous broken symmetry

Elaborating further on the above comments of P. W. Anderson, we refer to Aage Bohr Nobel lecture "Rotational motion in nuclei" [VIII]. Quoting (pp. 223–224):

> ... In a general theory of rotation, symmetry plays a central role. Indeed, the very occurrence of rotational degrees of freedom may be said to originate in a breaking of a rotational invariance, which introduces a deformation that makes it possible to specify an orientation of the system. Rotation represents the collective mode associated with such a spontaneous symmetry breaking (Goldstone mode).

[5]Although the equations describing the state of a natural system are symmetric, the majority of real physical systems with interactions between the particles tend to exhibit a lowest-energy state not having the full symmetries of the space or of the Hamiltonian describing the interaction [12].

...The Regge trajectories that have played a prominent role in the study of hadronic properties have features reminiscent of rotational spectra...

The condensates in superfluid systems involve a deformation of the field that creates the condensed boson or fermion pairs. Thus, the process of addition or removal of a correlated pair of electrons from a superconductor (as in a Josephson junction) or of a nucleon pair from a superfluid nucleus constitutes a rotational mode in the gauge space in which particle number plays the role of angular momentum (73). Such pair rotational spectra, involving families of states in different nuclei, appear as a prominent feature in the study of two-particle transfer process (74). The gauge space is often felt as a rather abstract construction but, in the particle-transfer process, it is experienced in a very real manner ...

(73) Anderson, P. W., Rev. Mod. **38**, 298 (1966).

(74) Middleton R. and Pullen D. J., Nuclear Physics. 51, 77 (1964); see also Broglia R. A., Hansen D., and Riedel C., Advances in Nuclear Physics **6**, 287, Plenum Press, New York, 1973.

The above quotation testifies to the breadth of Aage Bohr's perspective concerning the phenomenon of spontaneous symmetry breaking as embodied in the nuclear phenomena, and which encompasses condensed matter (Josephson effect), particle physics and field theory (Regge trajectory, Goldstone mode). Concerning the Goldstone mode, the following recount likely illuminates an important aspect of Aage Bohr's contributions to the broken symmetry issue. The apparently simple, almost bare style he used in the writings collected here, may induce the reader to take it at face value and, in this way, miss important facets. Let me illustrate this in the following example.

As stated throughout the Nobel lecture, the energy of rotational bands in nuclei depends quadratically on the variable conjugate to the angle (Euler, gauge, etc.) which becomes defined by the deformation associated with the spontaneous symmetry violation, i.e. angular momentum for quadrupole deformation in 3D space, number of particles N in the case of deformation in gauge space.

For simplicity, let us discuss rotations in the two-dimensional gauge space, within the framework of the single j-shell model. In this case, the BCS solution of the mean field pairing Hamiltonian can be carried out analytically and the ground state energy written as $U = (\hbar^2/2\mathcal{J})N^2, \mathcal{J}$ being the moment of inertia of the pairing rotational band, proportional to the number of Cooper pairs and inversely proportional to the pairing gap (Appendix C). Now, if this mode is to be interpreted as a Goldstone mode [14], it should approach zero as $N \to 0$, linearly in the number of particles. How does it do so quadratically? Because the above expression of U represents the energy of the system in the intrinsic, body-fixed frame of reference, where the nucleons are subject to the Coriolis force in gauge space $(-\lambda N)$. The linear asymptotic condition in N is operative in the laboratory system, where measurements can be carried out. In fact, $E_0 = \frac{\hbar^2}{2\mathcal{J}}N^2 + \lambda N$ fulfills the right asymptotic condition. This reasoning was implicitly contained in the answer: "you have to express the energy in the laboratory system", Aage Bohr gave to my question when I spoke with him on the subject.

I imagine, although I have no proof or confirmation of it, that the reason why a number if not all of Aage Bohr's contributions collected here are deceivingly simple and devoid of details, although rich in physical insight, is because he aimed at having the broad picture associated with the subject under discussion always present. Details, also important ones, could eventually be extracted from it by using footnotes, the literature and a few months of work. The more one learns and understands about the subjects in question, the more one finds that solutions and intuitions were there already, even if Aage Bohr could not possibly have known important details at that time.

3. Elementary modes of excitation. Pairing rotational and vibrational bands

On p. 30 of the January 1969 issue of the newly started magazine of science, "Scientific Research" [13] one reads: "The Institute's strong position in nuclear physics is further demonstrated by the work reported by [Aage] Bohr last summer at Trieste ([IV] RAB) and at Dubna ([V] RAB) in invited papers on excitations in nuclei and two-particle transfer reactions. Included in that work is a fascinating effort to elaborate a theoretical description of a mode of nuclear excitation involving nucleon-pair creation and annihilation. The phenomenon is analogous to the Josephson effect in superconducting metals, which arises from the tunnelling of pairs of electrons". In the introduction of [IV], Aage Bohr states:

> in ... systems where many degrees of freedom are involved, the central problem becomes that of establishing the relevant degrees of freedom and of developing the appropriate concepts that govern their interplay ... problems of fairly general nature that have puzzled nuclear physicists, and that might be suited to elicit response from the broader community of quantal physicists.

Present at the symposium were, among others, P. W. Anderson, J. R. Schrieffer, A. Abrikosov, A. De-Shalit, A. Salam, S. Weinberg, H. Feshbach, P. G. de Gennes, D. Pines and J. Schwinger.

The question of the identification of the degrees of freedom that are appropriate to describe the structure of nuclei was a central issue for Aage Bohr, and occupied a prominent place in his research and in conference presentations. Within this context and in lieu of many references, one can mention the LXIX Course of the International School of Physics "Enrico Fermi" he organized together with the editor in 1976 (see [IX]). About eighty participants among lecturers, observers and students from all parts of the world gathered to discuss, during the two weeks of the Conference that took place in the most congenial atmosphere of Villa Monastero at Lake Como, on the topic of Elementary Modes of Excitation in Nuclei, which was also the title of the Course.

Returning to [IV], one realizes that Aage Bohr refers to a definite, well-studied nucleus, in an attempt to make the discussion on the appropriate nuclear degrees of freedom to be as concrete as possible, and in doing so, he is able to make "universal" the associated physical concepts, and establish contact with condensed matter physicists, as well as field theory practitioners. Quoting:

> ...To introduce the quanta of nuclear physics, let us consider, as an example, the spectrum of elementary excitations based on the ground state of ^{208}Pb. This "vacuum" state has especially simple properties (has a

> minimum of degeneracy) on account of the closed-shell configuration of 82 protons and 126 neutrons.
>
> The elementary mode of excitation of fermion type ... can be simply interpreted on the basis of single-particle motion in a central potential with spin orbit coupling ... quasi-particles ... at the same time show renormalisation effects due to the coupling to the underlying ... particles embedded in the vacuum ...
>
> The boson excitations[6] ... can be associated with different types of deformations in the nuclear density and in the average potential.

As stated above, the selection of the ground state of the double closed shell nucleus $^{208}_{82}\mathrm{Pb}_{126}$ as the "vacuum" state, has a double hit effect: that to use a language readily understandable to the variety of field theory experts and, not less important, to indicate nuclear practitioners the possibility of attacking the nuclear many-body problem in terms of fermions (nucleons) and bosons (collective vibrations), making use of the powerful physical concepts and techniques which has produced Quantum Electrodynamics, let alone the modern picture of condensed matter, and of particle physics.

Once this is done he returns, as a conscientious artisan, to embody the grand design with concrete elements. Quoting again from [IV]:

> For example, the lowest states of ^{208}Pb correspond to shape oscillations of octupole... type. ... In addition to the quantum numbers λ and π (angular momentum and parity) one can characterise the excitations by additional ... quantum numbers ... The modes with nuclear transfer number[7] $\alpha = \pm 2$ may be viewed as oscillations in the nuclear pairing field, which represents the creation and annihilation of two nucleons at the same point in space or close together...

For a quantal physicist, the question at this point is how to be able to assess the validity of the picture and eventually appreciate its attractiveness. Simple. By looking at it. In other words, by carrying out measurements. Quoting again from [IV]:

> These transitions are now being studied by reactions such as (t, p), (^{3}He,n), etc., by which two nucleons are simultaneously exchanged between projectile and target, and the $\alpha = \pm 2$ modes ... are found to be excited with strongly enhanced matrix elements.

In fact, two-nucleon transfer reactions are the specific probe of pairing correlations in nuclei, in a similar way in which, as had been stated by P. W. Anderson in 1964 [14], and at the Trieste symposium [15], Cooper pair tunneling across a Josephson junction provides the tools to reveal superconducting (gauge phase) correlation. Aage Bohr presents, in half a page, the elementary modes of excitation picture of the atomic nucleus, setting special emphasis on pairing modes and on the specific probes to study them. He uses for the purpose a language which makes direct reference to techniques accessible to the broad many-body and field theoretical community. Certainly no small achievement.

[6] "composites of the fermions", Sec. 5 of [IV] (editor's comment).

[7] It is remarked that in [III] the symbol β was used to denote the same quantum number.

There is, however, something in the picture that turned out not as stated. It is correct that $\alpha = \pm 2$ modes are excited in two-nucleon transfer reactions with strongly enhanced matrix elements (cross-sections). However, it is neither because the pair is created and annihilated at the same point, nor because it is transferred simultaneously, but because the partner nucleons are (gauge) phase correlated over distances of the order of the correlation length ξ. Consequently, successive transfer can be operative without quasiparticle excitation. In fact, the main mechanism driving enhanced Cooper pair transfer in nuclei is successive transfer, and it leads to absolute cross-sections which are of the same order of magnitude as absolute single-particle transfer cross-sections, as experimentally observed.

Let us now quote in this connection from [V]:

> ... A two-neutron transfer reaction, by which two neutrons are added to ^{208}Pb in $J^\pi = 0^+$ state, produces the ground state of ^{210}Pb with a cross-section which is strongly enhanced as a consequence of the strong spatial correlation of the two neutrons in the ground state of ^{210}Pb. A similar enhancement characterises the (p,t) reaction leading to the ground state of ^{206}Pb. We may therefore view the ground state of ^{210}Pb and of ^{206}Pb as collective excitations of ^{208}Pb ... With the $\alpha = +2$ and $\alpha = -2$ quanta as building blocks, we can construct the pair vibrational spectrum ... the properties of the quanta are significantly affected by the ground-state correlations in ^{208}Pb, representing the virtual excitation of pairs of particles across the gap between the shells. In particular, this zero-point vibrational motion implies an enhancement of the two-particle transfer matrix elements. The effect is of similar nature to the enhancement of the E2 matrix elements for the low-energy quadrupole mode with $\alpha = 0$, resulting from the virtual excitation of quasiparticle pairs in the nuclear ground state.

While the above paragraph again sets in relation strongly enhanced two-nucleon transfer cross-sections with strong spatial correlation of the two neutrons, something which has turned out not as stated, it does provide nonetheless a masterful account of ground-state correlations — in other words, of vacuum zero point fluctuations of a quantal system displaying collective vibrations. Pairing vibrations in this case, and the positive feedback that fluctuations of two particle–two hole ($2p - 2h$) character have in the value of the two-particle transfer cross-sections. To clarify the importance and meaning of these poorly known (at that time) gauge space related phenomena, Aage Bohr sets them in connection with similar phenomena taking place in 3D-space when (particle–hole) $\alpha = 0$, quadrupole vibrations are excited by single-particle fields (inelastic scattering and Coulomb excitation processes) as well as in relation to the associated electromagnetic $E2$ decay. In this way, Aage Bohr makes use of a tradition going back to the Greeks ($\varphi\upsilon\sigma\kappa\alpha$, books of Aristotle which came before those of metaphysics) as expressed in simple terms by the French philosopher Simone Weil[8] "Toute théorie physique est fondée sur l'analogie qu'on établit entre des choses mal connues et des choses simples". This trait is characteristic of the selected papers published in the present volume, but also of the research of Aage Bohr in general. It includes, among others, parallels between atomic and nuclear physics, condensed matter and $N \to \infty$

[8] "All physical theories are based on the analogy which one establishes between things poorly known and simple things" Librarie PLON, Paris (1959).

concepts with those of strongly interacting finite many-body systems, but also and in a most profound way, between static and dynamic deformations in gauge and in 3D-space (in particular quadrupole deformations in this space). Within this context, ground state correlations resulting from the excitation of quasiparticle pairs of mainly $(p-h)$ character vibrations are closely connected with the random phase approximation (RPA) developed by Bohm and Pines,[9] more precisely, the backwards going amplitudes of the wavefunctions describing the collective modes. This is also the case for $2p-2h$ pairing vibrational modes. The fact that they are so much less known in the literature is in keeping with the fact that in condensed matter, paring vibrations play a negligible role with the exception made at $T = T_c$, or in the case of superconducting grains of dimensions d much smaller than the corresponding correlation length ($d \approx 200$ Å).

The role of pairing vibrations in nuclei competes on par with that of $p-h$ like vibrations, due to the finite size of the system and to the role spatial quantization plays in nuclei. Because both classes of (quasi) bosons are made out of the same particle and hole degrees of freedom, going beyond (harmonic) RPA, one finds processes in which the fermions participating in quadrupole and pairing zero point fluctuations are exchanged (exclusion principle) stabilizing the tendency ground state correlations have to induce quadrupole or superfluid phase transitions, the so-called pairing–quadrupole competition. This is also the reason why ground state correlations of multipole pairing vibrational modes increase two-nucleon transfer cross-sections but decrease inelastic or $B(E\lambda)$ processes. Conversely, ground state correlations associated with $(p-h)$ modes increase inelastic scattering cross-sections and $B(E\lambda)$ values, decreasing at the same time the associated two-nucleon transfer cross-sections. These topics played an important role in the definition of two-particle units developed to measure two-nucleon transfer cross-sections. Units which are similar to the single-particle, Weisskopf units used to quantify the strength of a $B(E\lambda)$ reduced the transition probability.

Returning to [IV]:

> ... The boson spectrum considered above has a vibrational character, i.e. it corresponds to oscillations of the various density components about the equilibrium represented by the vacuum state (the ground state of ^{208}Pb). However, for nuclei with many particles in unfilled shells, the independent particle motion gives rise to a highly degenerate ground state, and the system lifts the degeneracy by reducing the symmetry, i.e. by deforming itself (a well-known feature of quantal systems, referred to in molecules as the Jahn–Teller effect). In nuclei, the most important static deformations occur in the $\lambda^\pi = 2^+, \alpha = 0$ channel (quadrupole shape deformation) and in the $\lambda^\pi = 0^+, \alpha = \pm 2$ channels (monopole pairing deformation (superfluidity))... The pairing deformations provide a new dimension to the collective ... branches with baryon numbers A, A $\pm$ 2, A $\pm$4... representing rotational motion in gauge space ... One notes the close similarity to the description of the superfluid system, which Anderson[10] has discussed; the state with

[9]D. Bohm and D. Pines, Phys. Rev. **92**, 609 (1953). In connection to early applications of RPA to nuclear physics see: R. Arvieu and M. Vénéroni, Compt. Rend. **250**, 992, 2155 (1960), M. Baranger, Phys. Rev. **120**, 957 (1960), T. Marumori, Progr. Theor. Phys. (Kyoto) **24** 331 (1960), G. E. Brown, *Many-body Problems*, North Holland, Amsterdam (1972), T. J. Thouless *The Quantum Mechanics of Many-body Systems*, Academic Press, London (1972).

[10]See reference [15] (editor's comment).

> finite expectation of the pairing field is here the intrinsic state, and the phase ϕ is the collective coordinate, corresponding to the orientation of a deformed system. In the nucleus, the main effects are associated with the part of the phase field that is a constant over the nucleus ... The direct manifestation of the coherence is in the collective "rotational" spectrum, the numbers [members, RAB] of which are connected with large matrix elements for two-nucleon transfer ... In the heavy nuclei, it appears that the pairing is especially vulnerable to the effects of the rotation and, at the highest I-values studied, we are beginning to see the effects indicative of the disappearance of the stable pairing field. The effect is quite similar to the disappearance of superconductivity at the critical magnetic field ...
>
> Recently, Anderson objected to talking about phase transitions in the small particle superconductors[11] and it is of course true that we do not need non-analytic functions to describe the transitions in finite systems. On the other hand, the phenomena are closely related and, in a system like the nucleus, we have the possibility of studying the transition in terms of the individual quantal states. Thus, the transition from pair-correlated to normal systems with increasing angular momentum involves the coupling between the bands associated with the ground state and the excited bands representing fluctuations in the pairing field".

With the above quoted arguments, Aage Bohr asserts that at the basis of the unified description of the nucleus (shell and liquid drop models) one finds Landau's elementary modes of excitation, field concepts of renormalization and zero-point fluctuations of the nuclear vacuum typical of QED, and phase transitions subject to conspicuous quantal fluctuations.

In connection with his Trieste contributions [IV], he refers twice to Anderson's talk. Let us start with the second mention, concerning Anderson's objection to talking about phase transitions in the small particle superconductors: Aage Bohr argues that the two phenomena are closely related and he points out that in the nucleus, the transition from pair correlated to normal system as a function of the angular momentum can be studied in terms of individual states, in a detailed microscopic quantal description, namely, in terms of the coupling between the rotational states associated with the ground state and excited bands (Regge trajectories[12] mentioned also in [VIII]) whose intrinsic states are based on fluctuations of the pairing field.

It is suggestive that few years later, in his most influential 1972 article "More is different", Anderson states in connection with the many-body nuclear problem [10]: "It is only as the nucleus is considered to be a many-body system — in what is often called the $N \to \infty$

[11]Quoting from [15]: "... One of the most interesting limiting cases for the phenomenon of superfluidity is that of very small particles. For nine or ten years we have had one example of superconductivity in very small particles, namely the pairing phenomenon in nuclear matter. However, nuclei are not easy things to work with; we can apply the equivalent to H-fields by rotating them, but we cannot do thermal or coherence experiment on them, nor can we go to the macroscopic limit of nuclear matter. Thus it is nice that Giaever and co-workers have come up with a good series of experiments on tin particles in the interesting range $\leq 10^2$ Å...". We now know that experimental ingenuity has allowed for the study of the disappearance of pairing correlations as a function of nuclear temperature, and that superfluidity in the inner crust of neutron stars, a subject to which P. W. Anderson has contributed much, provides the $N \to \infty$ embodiment to nuclear matter superfluidity (editor's comment).

[12]Within this context see [16, 17, 18].

limit — that such a behaviour is rigorously definable. We say to ourselves: a macroscopic body of that shape would have such-and-such a spectrum of rotation and vibrational excitations completely different in nature from those which would characterise a featureless system. When we see such a spectrum, even not so separated, and somewhat imperfect, we recognize that the nucleus is, after all, not macroscopic: it is merely approaching macroscopic behavior". While no proof of the connection with Aage Bohr's remarks is available,[13] the above quote provides a rather accurate representation of Aage Bohr's ideas and an almost perfect picture of what the finite quantum nuclear many-body problem is all about.

In the first mention to Anderson in [IV], Aage Bohr refers to coherence in gauge space, in connection with pairing rotational bands, a subject which he discusses in even more detail in [V], in particular in connection with Fig. 3 of this paper, in reference to the Sn–isotopes. In the text, Aage Bohr mentions both the rotational band based on the condensate as intrinsic state, as well as on two-quasiparticle pairing vibrations. Also those in which the pair transfer goes into the next higher shell ($N > 82$) or the pair removal is associated with the lower shell ($N < 50$).

Systematic experimental data on Cooper pair pickup on the Sn-isotopes, namely $^{124-112}\text{Sn}(p,t)$ has allowed for a detailed probing of the validity of the pairing rotational picture, also based on low-lying two-quasiparticle 0^+ states (see [19] and refs. therein). The situation concerning the transfer modes to the higher (lower) shells, namely the giant pairing vibrations, is less clear. While the search of these modes in medium heavy nuclei has not been successful, recent work has provided clear indication of their presence in light nuclei (see [20, 21] and refs. therein, see also [22]).

Concerning the enhanced value of the absolute two-nucleon transfer cross-section associated with the population of collective pairing elementary modes of excitation, it has been found that successive transfer of nucleons dominates the process. Simultaneous transfer gives a modest contribution once orthogonalization of target and projectile wave functions is carried out. This result makes the process of addition and removal of a correlated pair of nucleons to a nucleus to be completely parallel to the process of addition and removal of a correlated pair of electrons between weakly coupled superconductors, that is the Josephson effect [23] (see also [24] and refs. therein, in particular, B. F. Bayman and J. Chen, *Phys. Rev. C* **26**, 1509 (1982)).

It is revealing of the subtlety of the phenomenon that the scientists, or more correctly geniuses, who best understood superconductivity in metals and superfluidity in nuclei, namely John Bardeen and Aage Bohr respectively, wrote, in connection with (electron, nucleon) Cooper pair tunneling

> In a recent note, Josephson...discuss the possibility of superfluid flow across the tunneling region, in which no quasiparticles are created. However,...pairing does not extend into the barrier, so there can be no such superfluid flow [25],

and:

[13] As already pointed out, at the time of the Trieste meeting, Anderson was interested in small metallic superconducting Sn particles, in connection with Giaever's experiments [15].

> The modes with nuclear number $\alpha = \pm 2$ may be viewed as oscillations in the pairing field, which represents the creation and annihilation of two nucleons at the same point in space or close together... These transitions are now being studied by reactions such as (t,p), $(^3\mathrm{H}, \mathrm{n})$, etc., by which two nucleons are simultaneously exchanged between projectile and target... [IV].

As elaborated in the remainder of this section, these statements did not turn out to be correct.[14] Two superconductors separated by an insulating barrier of thickness typically in the range 10–30 Å are said to be weakly coupled. One talks of a weak link, better known as Josephson junction. The probability of one electron to tunnel is small, let us say $P_1 \approx 10^{-10}$ [26]. Consequently, simultaneous transfer, with a probability $P_2 \approx (10^{-10})^2$ will not be observed. However, the fact that the wavefunctions of the electrons in the pair are phase–coherent $\left(U'_\nu + e^{-2i\phi} V'_\nu P^\dagger_\nu\right)|0\rangle$, where U'_ν and V'_ν are the BCS occupation amplitudes, ϕ the gauge angle and $P^\dagger_\nu = a^\dagger_\nu a^\dagger_{\bar{\nu}}$ creates, acting on the vacuum $|0\rangle$, a pair of electrons moving in time-reversal states, implies, according to Josephson, that one has to add the amplitudes before taking modulus square. That is, $P_2 = \lim_{\phi\to 0}\left|\frac{(U'_\nu\sqrt{P_1}+e^{-2i\phi}V'_\nu\sqrt{P_1})}{\sqrt{2}}\right|^2 \approx P_1$, $(U'_\nu V'_\nu \approx 1/2)$. In other words, and making use of Gorkov's derivation of the Ginzburg–Landau equations for BCS theory, $F(\mathbf{r},\mathbf{r}')$ is the amplitude for two electrons at $\mathbf{r}$ and $\mathbf{r}'$ to belong to a Cooper pair. The gap function $\Delta(\mathbf{r})$ is given by $V(\mathbf{r})F(\mathbf{r},\mathbf{r})$, where $V(\mathbf{r})$ is a local two-body interaction. In the insulating barrier $V(\mathbf{r})$ is zero. This does not imply the vanishing of F, the order parameter of the superconductor. The function $F(\mathbf{r},\mathbf{r}')$ can have large amplitudes for electrons separated by distances as large as the coherence length $(10^3 - 10^4$ Å), a quantity much larger than the width of the barrier. As a result, two electrons on the opposite sides of the barrier can be correlated equally well as when they both are on the same side of it. That is, without quasiparticle excitation [14, 27, 28].

In the nuclear case $F(\mathbf{r},\mathbf{r}')$ corresponds to the Cooper pair wavefunction $\Psi_0(\mathbf{r},\mathbf{r}') = \sum_{n',nlj} X_{nn'lj} R_{nl}(r) R_{n'l}(r') \left(\frac{2j+1}{2}\right)^{1/2} \frac{1}{4\pi} P_l(\cos\theta)$, where $R_{nl}(r)$ is the radial single particle wavefunction describing the state of quantum numbers nlj, X being the (normalized, Tamm–Dancoff) amplitudes and $P_l(\cos\theta)$ a Legendre polynomial of range l, θ being the angle between vectors $\mathbf{r}$ and $\mathbf{r}'$.

The correlation length of Cooper pairs in a superfluid nucleus, for example ^{120}Sn, is $\xi \approx 14$ fm, much larger than nuclear dimensions ($R_0 \approx 6$ fm). If the partner nucleon of a Cooper pair would be on average at a distance ξ, they could lower their relative momentum. Such a tendency is however frustrated by the fact that Cooper pairs are subject to the single-particle mean field, which acts as a very strong external field ($2\Delta/\epsilon_F \approx 0.07$). This field does not only force confinement. Because of spatial quantization with strong spin orbit effect it leads, among other things, to the presence of intruder states and thus parity mixing. Because the main pairing correlations take place at the nuclear surface, one can set $r = r' = R_0$. Assuming furthermore that all amplitudes are equal, one obtains $|\Psi(R_0, R_0, \theta)|^2 \sim$

[14] In the case of Cooper pair transfer in nuclei, like e.g. in (t,p) reactions, this view was supported by microscopic nuclear structure calculations (see e.g. G. F. Bertsch *et al.*, Nucl. Phys. **A91**, 123 (1967)) which, although quite useful to shed light on the many-body correlation aspects of nuclear Cooper pairs, did not do the same regarding the strong distortion this quasiboson suffered by the presence of the mean field (see below, discussion concerning the $F(\mathbf{r},\mathbf{r}')$ nuclear function).

$|\sum_l P_l(\cos\theta)|^2$. This relation leads to constructive (destructive) coherence as a function of parity, $\pi = (-1)^l$ for $\theta = 0°$ ($= 180°$), and thus to the result that fixing the position of one of the Cooper partners at the surface, the probability of finding the other one is centered around it (i.e. $\theta = 0$ situation). That is, inside the nucleus the partners of a Cooper pair approach each other as compared to pure two-particle configurations coupled to zero angular momentum.

In two-nucleon transfer reactions like e.g. ^{120}Sn$(p,t)^{118}$Sn(gs), the contact between projectile and target is weak. Nonetheless, even a very low density overlap can induce strong modifications in Cooper pairs. In particular, it allows nucleon partners of a Cooper pair to recede from each other — one neutron forming a virtual deuteron with the projectile (proton), the other neutron being at the antipode, one diameter apart ($2R_0 \approx 12$ fm), thus essentially at a correlation length from each other, but still very much pairing correlated. As a consequence, one nucleon can be transferred at a time through the action of the mean field and without quasiparticle excitation, successive being the dominant transfer mechanism. In other words, the "observable" in Cooper pair transfer reaction experiments is the formfactor[15] associated with the reaction process, e.g. the (p,t) reaction, which connects, within the framework of the distorted wave Born approximation (DWBA) the incoming proton waves with those of the outgoing triton. And the associated spatial distribution of the two neutrons of the Cooper pair is considerably more extended than that described by $|\Psi_0(\mathbf{r},\mathbf{r}')|^2$ and reflects, to a large extent, the effect of the mean field. Within this context one can posit that a Cooper pair tunneling between two nuclei in a successive transfer process becomes no more broken (quasiparticle excitation), than a single photon in a two-slit interference experiment. This is the reason why the absolute cross-section associated with Cooper pair transfer in pairing-correlated nuclei, has the same order of magnitude as that associated with one-nucleon transfer processes in the same system.[16]

4. On the particle-vibration coupling and nuclear field theory

At the basis of nuclear field theory (NFT) one finds the particle-vibration coupling (PVC).[17] And on the basis of this mechanism, one finds the question of the quantisation of angular momentum in heavy nuclei.

Quoting from [I]:

> ... while the extreme single particle model, in the case of odd nuclei, assumes the total angular momentum of the nucleus to be possessed by the single odd particle, the magnitude of the quadrupole moments, in certain cases, demands that at least 20 or 30 nucleons somehow share the nuclear angular momentum,

and continuing with a quotation from [II]:

---

[15]A two-nucleon formfactor is the non-local (i.e. a function of $\mathbf{r}_1, \mathbf{r}_2$) matrix element of $P^\dagger = \sum_{\nu>0} P^\dagger_\nu$ between initial and final nuclear states.

[16]For example [29] ^{10}Be(t,p)^{12}Be(gs) ($\sigma = 1.9 \pm 0.5$ mb, $4.4^0 \leq \theta_{cm} \leq 54.4^0$) as compared to [30] ^{10}Be(d,p)^{11}Be($1/2^+$) ($\sigma = 2.4 \pm 0.013$ mb, $5^o \leq \theta_{cm} \leq 39^o$) in the case of light nuclei around closed shell, and [31] ^{120}Sn(p,t)^{118}Sn(gs) ($\sigma = 3.024 \pm 0.907$ mb; $5^o \leq \theta_{cm} \leq 40^o$) as compared to [32] ^{120}Sn(d,p)^{121}Sn($7/2^+$) ($\sigma = 5.2 \pm 0.6$ mb; $2^o \leq \theta_{cm} \leq 58^o$).

[17]Within this context (PVC) see I. Hamamoto, Nucl. Phys. A **126**, 545 (1969); A**141**, 1 (1970).

> ... In the atomic case, the dynamic aspects of the field are of relatively minor significance, due to the stabilizing influence of the central nucleus. The attraction from the nucleus plays a dominant role in shaping the atomic field ... In the nucleus, however, there is no similar stabilizing agent and variations in the field associated with collective modes of motion play an essential role ... The motion of individual particles is strongly affected by the collective oscillations, and the nuclear properties are determined by the interplay of these two types of motion,

one arrives at the recognition that [VII]

> ... While the need to include both single-particle and collective degrees of freedom in the description of the nucleus was apparent, one faced problems arising from the fact that such a description employs an overcomplete set of variables.

Within this scenario we move to [IV], where after stating that:

> ...from the existence of the nuclear shell structure ... interactions can be represented in first approximation by an average field, and, thus, we may attempt to describe the coupling between the particles and the collective motion in terms of oscillations in the nuclear field associated with collective deformations. In this manner, the interactions are described by a three-field coupling 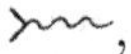 , representing the coupling of the particles to the nuclear deformation, rather than the four field coupling 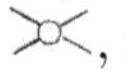 , representing the nucleonic interactions,[18]

one arrives at Sec. 5, namely the section on *Spurious degrees of freedom*, where we read:

> In the description of the nuclear dynamics in terms of collective and individual particle degrees of freedom, there arises the problem of the overcompleteness of the degrees of freedom, since the collective bosons are composites of the fermions. This is a general problem of many-body physics, but ... there is no cause for special worries, in that the elimination of the spurious degrees of freedom comes as part of the systematic analysis of the interaction effects.[19]

While Aage Bohr did not coauthor the paper where the NFT rules were presented [33], nor the closely related publication concerning the NFT treatment of the spurious states [36], his

[18] It turned out that both the three- and four-field couplings are to be considered in working out the interweaving of the basis states of NFT, namely single-particle and collective modes. The fact that a significant part of the second type of coupling is already included in generating the collective modes, implies that the NFT rules for evaluating the three- and four-point couplings involve a number of restrictions as compared with the rules that are employed when evaluating the original (four-point coupling) interaction [33].

[19] Within this context, one is reminded of the fact that Aage Bohr attended the 1948 Pocono Conference where Feynman presented his version of QED (after Schwinger). In explaining his formalism, Feynman pointed out that one did not need to worry about the Pauli principle in intermediate states, as there were diagrams which properly took care of it (see [34] p. 442, [35] p. 245).

contribution to both subjects which he felt to be closely related, was essential, as explained below, recounting developments which led to these publications.

4.1. *Aage Bohr and NFT*

In September 1972 RL and RAB had written[20] a short letter on the particle-vibration interpretation of the ^{209}Bi septuplet of states found around 2.6 MeV based on [37]. The main emphasis of the letter was placed on unraveling the interplay between pairing and surface vibrations. The importance of the pairing modes had been established in the case of ^{209}Pb [38]. This result had worked as the main source of inspiration in the analysis of the ^{210}Po(t,α) experiment carried out by RL and RAB [39]. In this paper, the importance of the proton pairing vibration modes was confirmed.[21]

During several rounds of discussions between RL and RAB with Aage Bohr, it became clear that the higher order corrections contained in the letter under preparation could neither be proved as the only ones operative, nor to be consistent with a systematic treatment of the particle-vibration mechanism. The challenge was taken up immediately.

In December 1972, RAB visited the group of Daniel Bés at the Atomic Energy Commission (AEC) of Argentina in Buenos Aires, and within a few days, many members of the group were deeply involved in the job of answering the questions of Aage Bohr, using his suggestions concerning open problems in connection with the particle-vibration coupling model, in particular, regarding energies, inelastic scattering and pick-up processes associated with the ^{209}Bi septuplet.[22] The first results emerged from the work of that month, where it was possible to determine a set of rules with which the particle-vibration model was able to give the right answer to various orders of perturbation in the particle-vibration coupling vertex.

Ben Mottelson eventually showed, utilizing some simple models that, in spite of these positive results, there were still some discrepancies with respect to the exact solution. With these models it was possible to individuate the key to the discrepancies, viz. the need to include the Hartree–Fock correction in the energy of the fermions.

In May 1973, DRB (visiting Copenhagen for an extended period of time), RL, RAB and BRM checked a variety of schematic models. It was found that the NFT rules (i.e. rules I-IV of Ref. [33]) gave the correct answer to any order of perturbation theory. Likely, the most detailed of all the controls carried out was that of the two-monopole phonon system, worked out explicitly up to tenth order of perturbation theory in the particle-vibration coupling vertices,[23] by DRB, RL and RAB. The corresponding graphs covered all the blackboards of the venerable and famous Auditorium A (see App. A). With this explicit result on hand, it

[20]RL: Roberto Liotta, RAB (editor)

[21]Within this context see also [40].

[22]The unification of structure and reactions (i.e. NFT (r+s)) was still far away (Ref. [9] p. 8 Foreword).

[23]In NFT there are two possible parameters upon which to expand in carrying out a perturbation expansion. The first is the strength of the interaction vertices (Λ)), i.e. the particle-vibration coupling vertices and the four-point vertices (V). The second parameter is $1/\Omega$ where $\Omega = \sum_j (j+1/2)$ is the effective pair degeneracy of the valence orbitals. The two types of parameters are in general connected through involved functionals. In schematic but still sensible physical models the connection is explicit, i.e. $\epsilon = O(1)$, $\Lambda = O(1/\sqrt{\Omega})$ and $V = O(1/\Omega)$. Another important feature is the number of internal lines which can be freely summed over. Each of them introduces a multiplicative factor Ω, i.e. $\sum$ (free internal line) $= O(\Omega)$. Within this scenario, the diagrams shown in Appendix A are of order $1/\Omega$, $1/\Omega^2$ and $1/\Omega^4$. As a rule, NFT calculations lead to converged results already in low order of perturbation theory, and provide an accurate, overall account of experimental findings in the case of realistic situations (see Refs. 42–44 as well as 38 of [9]).

was also possible to extend the sum of the perturbative series to infinite order. A manuscript presenting the nuclear field theory rules, together with selected examples from the schematic model was worked out. This was around the middle of 1973. Over a year of discussions and further calculations elapsed, in which Aage Bohr participated in an active way, before a final version was eventually written[24] and sent for publication[25] [33]. The acknowledgment recognizes a single source of inspiration: discussions with Aage Bohr.

The fact that NFT rules lead to the exact results in a variety of physically meaningful models, and that one could prove the equivalence between Feynman–Goldstone propagators involving only fermionic degrees of freedom and Feynman–NFT propagators involving fermionic and phonon degrees of freedom in the case of a general interaction, implied that NFT is corrected for the overcompleteness (redundant states), and for the violation of Pauli principle associated with the identity of the nucleons that implicitly participate in the collective modes and the particle degrees of freedom which appear explicitly (see Ref. 37 of [9]).

The question still remained of whether it was possible to see in a direct way how, within the framework of NFT, the redundant states were removed, a question which Aage Bohr had discussed with Rudolph Peierls in a visit to England. A model, the continuum model, was produced with the help of which one could show that the norm associated with spurious states was equal to zero, implying that they did not couple to the physical states. The model allowed for a graphical solution which made it possible to see, in the literal sense of the word, how redundant states were removed.

Aage Bohr checked the results of the model, concentrating his attention on the roots associated with the spurious states ($|Z| = 0$, App. B), as well as on rule (I) of NFT which states that one should start and end diagrams making use of allowed states. In the process of further discussions and controls, almost a year elapsed before the final version of the letter was ready and eventually sent for publication [36].

In the meantime DRB, RL and RAB had taken up again the challenge presented by the ^{209}Bi multiplet, this time the group having been reinforced by the arrival of Pier Francesco Bortignon. A year later, this effort resulted in a systematic presentation of NFT, again with innumerable contributions from Aage Bohr (see Ref. 42 of [9]).

5. The elements of creativity

In March 1971 the Institute celebrated the 50th anniversary of its inauguration (3/3/1921). On that occasion, Aage Bohr opened his speech [X]:

> ... cordially welcome all our guests, who have come from close and distant places...

And again, in the final part of his talk he expresses:

> Our thoughts go further to the large group of collaborators, close to a thousand, from all parts of the world who, during the times, have taken part in the Institute's activities, and provided much inspiration. We would

[24]The Copenhagen–Buenos Aires nuclear field theory had been born.

[25]In parallel, and on the basis of these developments, DRB and his group, together with RAB had produced a series of papers which were eventually published (see Refs. 33–37 of [9]).

> have liked to use this opportunity to have as many of them as possible to discuss the actual problems, but the room capacity of the Institute did not allow for it, and a meeting with so many participants would have hardly allowed for discussions to be carried out in the way we are accustomed to do ... We can, in any case, rejoice with regards to the lively contact still existing with this group of former collaborators, and imagine that all of them feel themselves as part of one large family.

Another recurrent theme of his speech concerned the fact that the

> Interaction between theory and experiment, ... has all the time occupied a central role ...

Of course the above quotes can hardly do justice to the unique creative atmosphere of the Institute in the 1960's and 1970's. Visitors not only had access to contribute firsthand to important projects, but also to become responsible for tasks and scopes which, to be developed, induced and helped them to absorb most of what was known on the subject at a surprising pace. This was real, generous, totally open scientific collaboration. Who thought or did this or that was irrelevant. Or at least that was what RAB felt. You got a result and reported on it right away, leaving also your handwritten notes. One was participating in a unique enterprise, which implied the uncovering of marvelous, unexplored venues. This also took place in the weekly experimental group meetings, where the latest measurements, many times not yet fully analyzed, were confronted with theories and models in the making.

Collaboration for Aage Bohr had the meaning of interplay of different individuals' competencies and sensibilities within an extended family, a tradition which goes very much back to his father. Visitors and Institute members alike tried to live up to the science, but also to the deeply human, generous view of life as a whole that was natural for Aage Bohr. Within this scenario it is fair to say that one was attracted, likely mesmerised, by the never-ceasing emergence of new ideas and results being produced at the Institute, admiring Aage Bohr's leadership this side of idolatry.

This was one — likely the central — element at the basis of creativity during the 1960s and 1970s at Blegdamsvej, and of the fact that, at that time, the Institute was a Mecca for low-energy nuclear physics. But equally unique in the 1960s was Denmark in general, and Copenhagen in particular. A somewhat isolated but nonetheless cosmopolitan city, model of social concern and democracy. People from all over the world traveled to Copenhagen to learn how to do physics in a great school and in a most congenial social atmosphere.

This was also one of the reasons why, at the Institute, one was able to have visitors from eastern Europe, in particular, from East Germany, during the time of the Berlin wall, and from the Soviet Union in spite of the iron curtain. And at the same time young Americans, those coming fresh from their Ph.D defence, but also those who invoked research work at the Institute to avoid being drafted to the war in Vietnam.

About halfway in his March 3, 1972 talk, Aage Bohr refers to the efforts of his father in connection with an open world between nations, in particular, concerning the scientific endeavour. To such recollection, he expresses his view as Director of The Niels Bohr Institute:

> For the Institute, with its position within the international scientific cooperation, it is a natural duty to contribute to the development in this direction through a further build up of such collaborations.

Aage Bohr rounds up his talk by stating that his father used to say, in reference to the concept of fortune, that it describes a situation in which things go better for one than one deserves, and concludes:

> Let us hope that on fifty years one will be able to say, that fortune was with us.

We are almost there. The insight developed and the novel concepts unveiled by Aage Bohr concerning the physics of many-body systems at large and of the atomic nucleus in particular, have become common knowledge of practitioners. They have reached the four corners of the world, helping to discover new, unexpected facets of nuclear structure, as well as parallels with other fields of research,[26] besides the fact that their importance was recognised in 1975 by the Royal Swedish Academy of Sciences. One can then posit that, in fact, one "had" fortune, and that "one" was not only Aage Bohr, but also the whole of the nuclear physics community.

Acknowledgments

Discussions with and suggestions of Pier Francesco Bortignon[27] have played a central role in the content and in the form of the above notes. His advice, as always in our long collaboration, has been a source of inspiration and a point of reference. Illuminating discussions with Christian Joas on many interdisciplinary aspects lying at the borderline between many-body physics research and history of physics are here acknowledged. Important input from Enrico Vigezzi concerning a number of central technical and general physical issues are gratefully recognized. I am also indebted to Roberto Liotta for comments on the manuscript. A special thanks to Daniel Bès for his illuminating remarks concerning both the general layout as well as for specific technical points. I am beholden to Robert J. Sunderland for the translation of paper [X].

References

[1] J. Schwinger, *Quantum Mechanics*, Springer, Berlin (2003), Prologue p. 23.
[2] Workshop on "Challenges in the investigation of double charge-exchange nuclear reactions: towards neutrino-less double beta decay" (NUMEN2015); Conference on neutrino and nuclear physics (CNNP2017).
[3] D. R. Bes and R. A. Sorensen, *The paring-plus-quadrupole model*, Adv. in Nucl. Phys., Vol. 2 (1969) p. 129.

[26]See e.g. *Fifty Years of Nuclear BCS, Pairing in Finite Systems*, Eds. R. A. Broglia and V. Zelevinsky, World Scientific, Singapore (2013), which contains 46 contributions filling 670 pages penned by leading specialists around the world. Also the Royal Swedish Academy of Science publication of the Focus issue to celebrate the 40-year anniversary of the 1975 Nobel Prize, Ed. J. Dudek, Physica Scripta Vol. **91** (2016), containing 51 contributions from some of the most active groups and researchers in the field of nuclear physics.
[27]Deceased August 27th 2018.

[4] J. Ungrin, R. M. Diamond, P. O. Tjøm and B. Elbek, *Inelastic deuteron scattering in the lead region*, Kong. Dan. Videns. Selsk. Mat.-Fys. Medd. **38**, 8 (1971).
[5] D. R. Bes, *Beta-vibrations in even nuclei*, Nucl. Phys. **49**, 544 (1963).
[6] R. Arvieu *et al.*, *Description de noyaux $Z = 50$ par un modele de quasi-particules en interaction*, Phys. Lett. **4**, 119 (1963).
[7] B. Elbek, *Determination of nuclear transition probabilities by Coulomb excitation*, Ejnar Munskgaard, Copenhagen (1963).
[8] D. R. Bes, *The field treatment of the nuclear spectrum. Historical foundation and two contributions to its ensuing development*, Phys. Scripta **91**, 063010 (2016).
[9] R. A. Broglia, P. F. Bortignon, F. Barranco, E. Vigezzi, A. Idini and G. Potel, *Unified description of structure and reactions: implementing the nuclear field theory program*, Phys. Scripta **91**, 063012 (2016).
[10] P. W. Anderson, *More is different*, Science **177**, 393 (1972).
[11] P. W. Anderson, *A helping hand on elementary matters*, Nature **405**, 736 (2000).
[12] P. W. Anderson, *Basic Notions of Condensed Matter Physics*, Addison Wesley, Reading, Massachussetts (1984).
[13] H. L. Davis, *Jet-age mecca for international physics*, Sci. Res. **4**, 27 (1969).
[14] P. W. Anderson, *Special effects in superconductivity*, in The Many-Body Problem, Vol. 2, E. R. Caianello, editor. Academic Press, New York, (1964) 113.
[15] P. W. Anderson, *Macroscopic coherence and superfluidity*, in Contemporary Physics: Trieste Symposium 1968, Vol. I, IAEA, Vienna (1969) p. 47.
[16] A. Johnson, H. Ryde and S. A. Hjorth, *Nuclear moment of inertia at high rotational frequencies*, Nucl. Phys. A**179**, 753 (1972).
[17] F. S. Stephens and R. S. Simon, *Coriolis effects in the yrast states*, Nucl. Phys. A**183**, 257 (1972).
[18] A. Molinari and T. Regge, *Rotational motion at large angular momentum in even-even deformed nuclei*, Phys. Lett. B**41**, 93 (1972).
[19] G. Potel *et al.*, *Quantitative study of coherent pairing modes with two-neutron transfer: Sn isotopes*, Phys. Rev. C **87**, 054321 (2013).
[20] F. Cappuzzello *et al.*, *Signatures of giant pairing vibration in the ^{14}C and ^{15}C atomic nuclei*, Nature Commun. **6**, 6743 (2015).
[21] P. F. Bortignon *et al.*, *Elastic response of the atomic nucleus in gauge space: giant pairing vibrations*, Eur. Phys. J. A**52**, 280 (2016).
[22] M. Laskin, R. F. Casten, A. O. Macchiavelli, R. M. Clark, and D. Bucurescu, *On the population of the giant pairing vibration*, Phys. Rev. C**93**, 034321 (2016).
[23] B. D. Josephson, *Possible new effects in superconductive tunnelling*, Phys. Lett. **1**, 251 (1962).
[24] G. Potel *et al.*, *Cooper pair transfer in nuclei*, Rep. Prog. Phys. **76**, 106301 (2013).
[25] J. Bardeen, *Tunneling Into Superconductors*, Phys. Rev. Lett. **9**, 147 (1962).
[26] A. B. Pippard, *The historical context of Josephson's discovery*, in 100 years of Superconductivity, Eds. H. Rog and P. H. Kes, CRC Press, Boca Raton, FL (2012) p. 29.
[27] P. W. Anderson, *How Josephson discovered his effect*, Phys. Today, November (1970) p. 23.
[28] D. G. McDonald, *The Nobel laureate versus the graduate student*, Phys. Today, July (2001) 46.
[29] H. T. Fortune, G.-B. Liu and D. E. Alburger, *$(sd)^2$ states in* ^{12}Be, Phys. Rev. C **50**, 1355 (1994).
[30] K. T. Schmitt *et al.*, *Reactions of a ^{10}Be beam on proton and deuteron targets*, Phys. Rev. C **88**, 064612 (2013).
[31] Bassani *et al.*, (p,t) *Ground-State* $L = 0$ *Transitions in the Even Isotopes of Sn and Cd at 40 MeV, $N = 62$ to 64*, Phys. Rev. B **139**, 830 (1965).

[32] M. J. Bechara and O. Dietzch, *States in* ^{121}Sn *from the* ^{120}Sn$(d,p)^{121}$Sn *reaction at 17 MeV*, Phys. Rev. C **12**, 90 (1975).
[33] D. R. Bes, G. G. Dussel, R. A. Broglia, R. Liotta and B. R. Mottelson, *Nuclear field theory as a method of treating the problem of overcompleteness in descriptions involving elementary modes of both quasi-particles and collective type*, Phys. Lett. **52B**, 253 (1974).
[34] S. S. Schweber, *QED and the Men who Made It*, Princeton University Press, Princeton (1994).
[35] J. Mehra, *The Beat of a Different Drum*, Clarendon Press, Oxford (1996).
[36] R. A. Broglia, B. R. Mottelson, D. R. Bes, R. Liotta and H. M. Sofia, *Treatment of the spurious states in nuclear field theory*, Phys. Lett. **64B**, 29 (1976).
[37] B. R. Mottelson, *Properties of individual levels and nuclear models*, in Proc. Intern. Conf. on Nuclear Structure, Tokyo 1976, Ed. J. Sanada, J. Phys. Soc. Jap. **24**, 87 (1968).
[38] D. R. Bes and R. A. Broglia, *Effect of the Multipole Pairing and Particle-Hole Fields in the Particle-Vibration Coupling Problem of* ^{209}*Pb. (I)*, Phys. Rev. **C3**, 2349 (1971).
[39] P. D. Barnes, E. Romberg, C. Ellegaard, R. F. Casten, O. Hansen, T. J. Mulligan, R. A. Broglia and R. Liotta, *Proton-hole state in* ^{209}*Bi from the* ^{210}*Pb (t,α) reaction*, Nucl. Phys. **A195**, 146 (1972).
[40] P. F. Bortignon, R. A. Broglia, D. R. Bes, R. Liotta and V. Paar, *On the role of the pairing modes in the* $(h_{9/2} \otimes 3^-)$ *multiplet of* ^{209}*Bi*, Phys. Lett. **64B**, 24 (1976).

Appendix A

Examples of fourth, eighth and tenth-order (in the particle-vibration coupling vertex Λ) NFT diagrams describing the interaction between two monopole particle-hole vibrational modes. The particles (holes) are described by arrowed lines pointing upwards (downwards), while the wavy lines correspond to the vibrational modes. Time is assumed to run upwards. No diagrams involving ground state correlations, i.e. diagrams containing processes where all three lines are connected with a vertex coming from above or below it are shown. They can be obtained from time ordering of the processes displayed. On page xxxi diagrams are shown containing a single independent summation and four particle-vibration coupling vertices. Consequently, they are of order $(1/\sqrt{\Omega})^4\Omega = 1/\Omega$ (see footnote 23). Diagrams shown on the right lowest corner of page xxxii (double line inset) contain two free summations and $n(m)$ (8(0), 6(1), 4(2)) particle-vibration (four-point) vertices. Consequently they are of order $(1/\sqrt{\Omega})^n \times (1/\Omega)^m \times \Omega^2 = 1/\Omega^2$. The rest of the diagrams on page xxxii contain a single free summation and 10(0), 8(1), 6(2), 4(3) particle-vibration (four point) vertices. Consequently, they are of order $1/\Omega^4$. For details see Ref. 35 of [9].

Counting the graphs

Example 4th order.

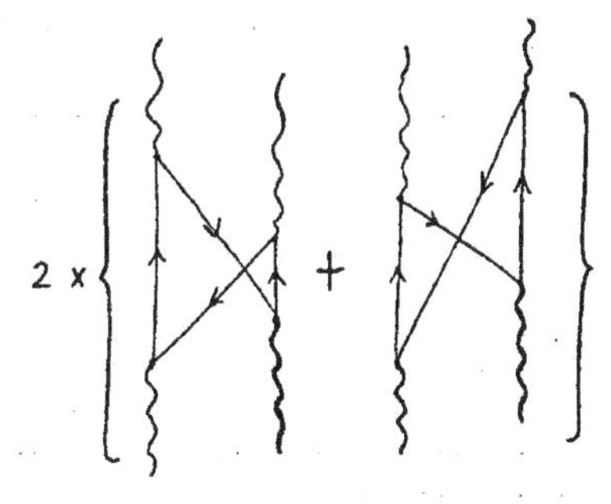

All other structures are topologically equivalent. The factor 2 comes because of the presence of two identical bosons in initial and final states, i.e.

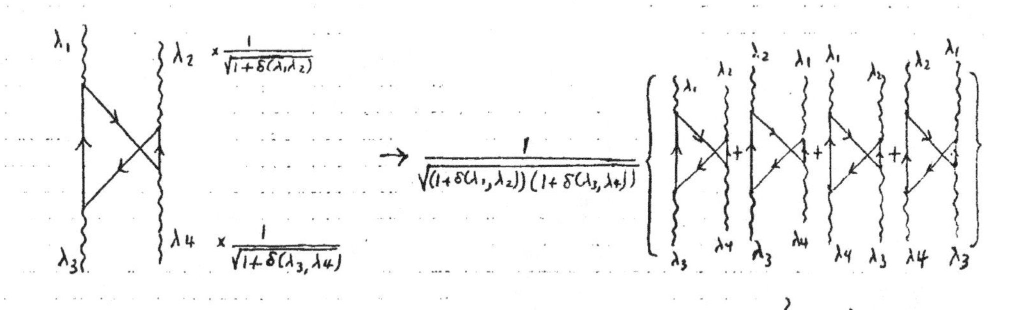

For the case $\lambda_1=\lambda_2=\lambda_3=\lambda_4$ we obtain $2\times$

Order of the different dynamical parameters

The quantity to be used as the expansion parameter is $\frac{1}{\Omega}$ (as in the Beliaev-Zelevinsky method). The other parameter $X=\frac{2V\Omega}{\varepsilon}$ is assumed to be relatively small in such a way that we are all the time far from the phase transition.

$$X=\frac{2V\Omega}{\varepsilon}\ ;\ X=\mathcal{O}(1)\quad \text{thus}\quad V=\mathcal{O}\left(\frac{1}{\Omega}\right)$$

$$\varepsilon'-\omega=\mathcal{O}(1)$$

$$V=\mathcal{O}\left(\frac{1}{\Omega}\right)$$ ← (this result was expected as to be far from phase transition ($X\ll1$) the int. V should be weak)

$$\Lambda=\mathcal{O}\left(\frac{1}{\sqrt{\Omega}}\right)$$

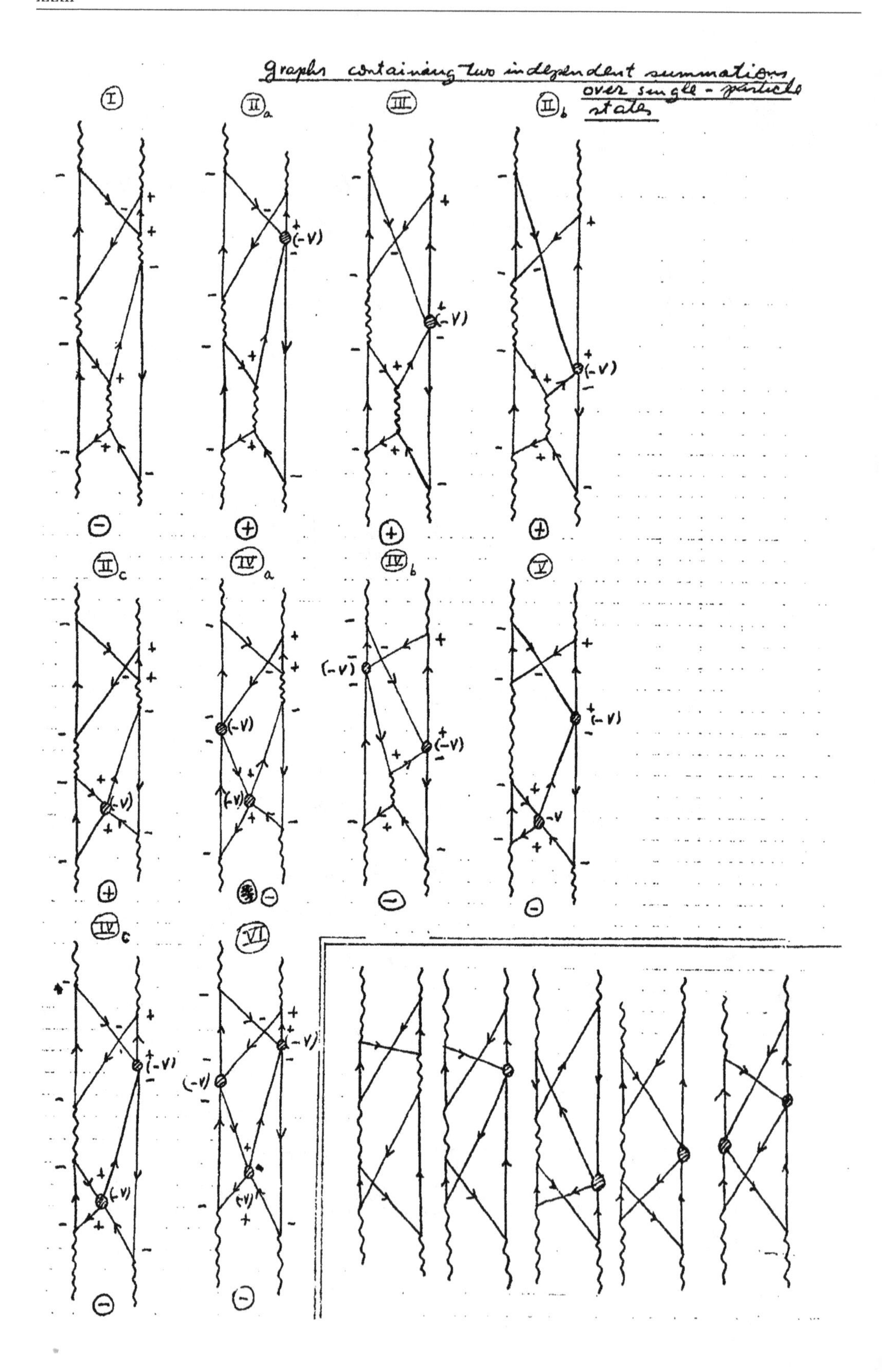
graphs containing two independent summations over single-particle states
I
II a
III
II b
II c
IV a
IV b
V
IV c
VI
(-V)
-V

Appendix B

Notes dated 28-9-1975 and 4-10-1975, Aage Bohr gave RAB with the derivation of the solution of the continuous model at the basis of [36], in which special attention is paid to the $|Z| = 0$ roots (spurious states).

28-9-75.

I.) Determination of normal modes i

eqs $\left| (\varepsilon_m - \omega_i)\,\delta(m,m') - V \right| = 0$

or $$\frac{1}{V} = \sum_m \frac{1}{\varepsilon_m - \omega_i} \qquad i = 1, 2, \ldots 2\Omega.$$

wave f.

$$\langle m\text{-}m \mid i \rangle = \frac{\Lambda_i}{\varepsilon_m - \omega_i} \qquad \Lambda_i = \left[\sum_m \frac{1}{(\varepsilon_m - \omega_i)^2} \right]^{-1/2}$$

II.) Treatment of states i, m ; ~~states with a~~ these states form overcomplete, non-orthogonal set.

We treat ~~them~~ interaction in terms of diagrams and get

$$H_{ii'} = - \frac{\Lambda_i \Lambda_{i'}}{E - \varepsilon_m - V} \quad \text{depending on energy.}$$

Diagonalization leads to

$$\left| (\omega_i - E_q)\,\delta(i,i') - H_{ii'} \right| = 0 \qquad (\text{note } E - \varepsilon_m - V \text{ constant, indep. of } i)$$

or $$E_q - \varepsilon_m - V = \sum_i \frac{\Lambda_i^2}{\omega_i - E_{qm}}$$

having $2\Omega - 1$ "regular roots" and $E_q = \varepsilon_m$ as "spurious" double root.

"wave function" is

$$\langle i \mid a_m \mid q,m \rangle = \xi_{imq} = - N_{qm} \cdot \frac{\Lambda_i}{\omega_i - E_{q,m}}$$

where $N_{q,m}$ ~~can be obtained from~~ is normalization factor. In determination of $N_{q,m}$, one must take into account non-orthogonality of states $a^\dagger_m|i\rangle$

$$\langle i' \mid a_m a^\dagger_m \mid i \rangle = \sum_{m' \neq m} \frac{\Lambda_i \Lambda_{i'}}{(\varepsilon_{m'} - \omega_i)(\varepsilon_{m'} - \omega_{i'})} = \delta(i,i') - \frac{\Lambda_i \Lambda_{i'}}{(\varepsilon_m - \omega_i)(\varepsilon_m - \omega_{i'})}$$

then

Hence

$$N_{qm}^2 \cdot \sum_{i,i'} \frac{\Lambda_i}{\omega_i - E_q} \frac{\Lambda_{i'}}{\omega_{i'} - E_q} \left(\delta(i,i') - \frac{\Lambda_i \Lambda_{i'}}{(\varepsilon_m - \omega_i)(\varepsilon_m - \omega_{i'})} \right) = 1 .$$

or since

$$\sum_i \frac{\Lambda_i^2}{(\omega_i - E_q)(\varepsilon_m - \omega_i)} = \sum_i \frac{\Lambda_i^2}{\varepsilon_m - E_q} \left(\frac{1}{\omega_i - E_q} + \frac{1}{\varepsilon_m - \omega_i} \right)$$

also root

$$= \frac{1}{\varepsilon_m - E_q} (E_q - \varepsilon_m - V + V) = -1 .$$

$$\to N_{qm}^2 \left(\sum_i \frac{\Lambda_i^2}{(\omega_i - E_q)^2} - 1 \right) = 1 .$$

and orthogonality follows from

$$\langle q'm | qm \rangle = N_{qm} N_{q'm} \sum_{i,i'} \frac{\Lambda_i}{\omega_i - E_q} \frac{\Lambda_{i'}}{\omega_{i'} - E_{q'}} \left(\delta(i,i') - \frac{\Lambda_i \Lambda_{i'}}{(\varepsilon_m - \omega_i)(\varepsilon_m - \omega_{i'})} \right)$$

$$= N_{qm} N_{q'm} \left(\sum_i \frac{\Lambda_i^2}{E_q - E_{q'}} \left(\frac{1}{\omega_i - E_q} - \frac{1}{\omega_i - E_{q'}} \right) - 1 \right) = 0 .$$

$$\frac{1}{E_q - E_{q'}} (E_q - \varepsilon_m - V - E_{q'} - \varepsilon_m - V)$$

~~Now, the state~~

The states q, m can be written $\quad |q, m\rangle = a_m^{\dagger} \cdot \sum_i \xi_{i,mq} |i\rangle .$

Now, the state with $E_q = \varepsilon_m$ ~~gives~~ is given by

$$\sum_i \xi_{i,mq} |i\rangle = N_{q,m} \cdot \sum \frac{\Lambda_i}{\omega_i - E_{qm}} |i\rangle = -N_{q,m} |m, -m\rangle .$$

It is this the spurious state that vanishes (cannot be normalized) while the $2\Omega - 1$ additional roots are orthogonal, and free of spuriosity.

4-10-75

III.) One might ask whether correct to diagonalize the effective H though the space is not orthogonal (has a redundance).

However, diagonaliz. same whether or not orthogonal; in fact, if $\Sigma(H_{ii'} - E\delta_{ii'})\, \xi_{i'} = 0$, we surely have stationary state.

This condition might not be necessary (though sufficient).

~~If~~ The necessary condition is that

$$\sum_{i''} Z_{ii''} \sum_{i'} \left(H_{i''i'} - E\,\delta_{i''i'}\right) \xi_{i'} = 0$$

where
$$Z_{ii''} = \langle i'' | a_m a_m^\dagger | i \rangle = \sum_{m'\neq m}' \frac{\Lambda_i \Lambda_i''}{(\varepsilon_m' - \omega_i)(\varepsilon_m' - \omega_i'')}$$ is overlap

or

$$|Z\,(H - E)| = 0$$

giving either $|(H-E)| = 0$

or $|Z| = 0$

last equation has only one root (since space is overcomplete one) corresponding to spurious state

$$\sum_i \xi_i(\text{spurious}) \cdot Z_{ii'} \propto \sum_i' \frac{\Lambda_i}{\omega_i - \varepsilon_m} \sum_{m'\neq m} \frac{\Lambda_i}{(\varepsilon_m' - \omega_i)} \cdot \frac{\Lambda_i'}{(\varepsilon_m' - \omega_i'')} = 0$$

since $$\sum_i' \frac{\Lambda_i^2}{(\omega_i - \varepsilon_m)(\omega_i - \varepsilon_m')} = \langle m, -m | m', -m' \rangle = \delta(m, m').$$

But this root already included in $|(H-E)| = 0$ (in fact, as double root).

Appendix C

We consider N nucleons moving in a single j-shell of energy ϵ_j and total angular momentum $(ls)j$. The number of pairs moving in time-reversal states which can be accommodated in the shell is,

$$\Omega = \frac{2j+1}{2}. \tag{1}$$

The BCS occupation factors are in this model

$$V = \sqrt{\frac{N}{2\Omega}} \quad , \quad U = \sqrt{1 - \frac{N}{2\Omega}}, \tag{2}$$

the number of Cooper pairs being

$$\alpha_0 = \Omega UV = \frac{\sqrt{N(2\Omega - N)}}{2}, \tag{3}$$

as can be seen by inspection setting e.g. $N = \Omega$, in which case $\alpha_0 = \Omega/2$ as expected. The solutions of the number and gap equations associated with the BCS diagonalization of a pairing force with constant matrix elements G are

$$\lambda = -\frac{G}{2}(\Omega - N), \tag{4}$$

and

$$\Delta = \frac{G}{2}\sqrt{N(2\Omega - N)} = G\alpha_0 \tag{5}$$

while the associated ground state energy in the laboratory system is,

$$E_0 = (E_0)_{\text{lab}} = (E_0)_{\text{intr}} + \lambda N = U + \lambda N = -\frac{\Delta^2}{G} = -\frac{G}{2}N\Omega + \frac{\hbar^2}{2\mathcal{J}}N^2. \tag{6}$$

In writing the above equation, the assumption has been made that $\epsilon_j = 0$. Making the ansatz $N << \Omega$,

$$\lambda = -\frac{G\Omega}{2}, \tag{7}$$

and

$$E_0 = \lambda N + \frac{\hbar^2}{2\mathcal{J}}N^2, \tag{8}$$

the quantity

$$\mathcal{J} = \frac{2\hbar^2}{G} = \frac{2\hbar^2}{\Delta}\alpha_0, \tag{9}$$

being the moment of inertia of the rotational band in gauge space.

PHYSICAL REVIEW VOLUME 81, NUMBER 1 JANUARY 1, 1951

On the Quantization of Angular Momenta in Heavy Nuclei

AAGE BOHR
*Department of Physics, Columbia University, New York, New York**
(Received May 31, 1950)

The individual particle model of nuclear structure fails to account for the observed large nuclear quadrupole moments. It is possible, however, to allow for the existence of the quadrupole moments, and still retain the essential features of the individual particle model, by assuming the average field in which the nucleons move to deviate from spherical symmetry. The assumptions underlying such an asymmetric nuclear model are discussed; this model implies, in particular, a quantization of angular momenta in analogy with molecular structure. The asymmetric model appears to account better than the extreme single particle model for empirical data regarding nuclear magnetic moments.

I. INTRODUCTION

THE individual particle model, which describes the stationary state of a nucleus in terms of the motion of the individual nucleons in an average nuclear field, has accounted successfully for a large number of nuclear properties.[1] In the simplest form of this model the nucleons are assumed to move in a field of spherical symmetry and the quantization of angular momenta is similar as in atomic structures.

This extreme model meets with the difficulty, however, that nuclei are found to have very large electric quadrupole moments. Even in the case of nuclei with one particle more or one particle less than the number required for closed shells, the quadrupole moments are, in general, several times larger than can reasonably be attributed to a single particle.[2]

This circumstance has a direct bearing on the problem of the quantization of angular momenta in the nucleus, quite apart from what other modifications of the model may be necessary. In fact, while the extreme single particle model, in the case of odd nuclei, assumes the total angular momentum of the nucleus to be possessed by the single odd particle, the magnitude of the quadrupole moments, in certain cases, demands that at least 20 or 30 nucleons somehow share the nuclear angular momentum.

Little is known regarding the coupling of angular momenta in heavy nuclei, and we may have to do with complicated coupling schemes. However, it is possible to allow for the existence of the large quadrupole moments, and still retain many essential features of the individual particle model by assuming that the average nuclear field in which the nucleons move deviates from spherical symmetry.

This model involves certain assumptions regarding the dynamical properties of the nucleus and has a number of direct implications regarding nuclear moments which form the subject of the present paper.

A model of this type has recently been considered independently by J. Rainwater,[3] who found it possible to account for the order of magnitude of the quadrupole moments by estimating the nuclear deformation produced by the centrifugal pressure of the odd particle. The following considerations, however, are largely independent of the origin of the nuclear deformations responsible for the large quadrupole moments.

II. THE ASYMMETRIC NUCLEAR MODEL

The model of the asymmetric nucleus suggests an analogy with molecular structure, in which the electrons move in average force fields which deviate from spherical symmetry. As is well known, the possibility of a simplified treatment of molecular states is based on the fact that the masses of the nuclei are very large compared with the mass of the electron. Consequently the motion of the nuclei, their vibration and rotation, only adiabatically influences the electronic motion, whose frequencies are very large compared to those characteristic of the nuclear motion.

The model of a nucleon moving in an average field of a deformed nucleus assumes, in a similar manner, that the shape and orientation of the nucleus is approximately fixed over time intervals comparable with the periods of the single particle motion. In contrast with the molecular case, there are here no heavy particles to provide the necessary rigidity of the structure. However, nuclear matter appears to have some of the properties of coherent matter which makes it capable of types of motion for which the effective mass is large as compared with the mass of a single nucleon.

Thus it is generally assumed that nuclei may vibrate under the influence of surface tension, and since such oscillations involve the whole nuclear mass, the shape of the nucleus will change very slowly compared with the single particle motion. Moreover, the magnitude of the quadrupole moments implies that in general a large number of nucleons are involved in the deformation of the nucleus, and therefore suggests that the frequencies associated with the rotation of the deformed nucleus

* On leave from the Institute for Theoretical Physics, Copenhagen, Denmark.

[1] M. G. Mayer, Phys. Rev. **78**, 16, 22 (1950). These articles contain references to previous literature regarding the individual particle model. E. Feenberg, Phys. Rev. **77**, 771 (1950).

[2] Townes, Foley, and Low, Phys. Rev. **76**, 1415 (1949).

[3] J. Rainwater, Phys. Rev. **79**, 432 (1950). I am indebted to Dr. Rainwater for informing me of his results prior to publication.

may be considered to be small compared with single particle frequencies.

It may be noted that the nucleus is not assumed to rotate like a rigid structure; this assumption would imply the existence of very closely spaced rotational energy levels in heavy nuclei, and such levels have not been observed. It seems quite possible that a rotation of the nucleus involves the motion of only a fraction of the nuclear particles, due to an incomplete rigidity of the nuclear structure, or due to exchange effects of the type considered by Teller and Wheeler.[4] The effective moment of inertia to be associated with the rotation may thus be considerably smaller than that of the whole nucleus. The rotational level spacing, although small compared with that of a single nucleon, may therefore still be compatible with the absence of very low lying nuclear excited states.

In the analysis of nuclear stationary states we may now first consider the motion of a nucleon in the approximately fixed field produced by the core of the nucleus. In a non-central field the angular momentum of the particle will not be a constant, but for cylindrical symmetry will precess around the symmetry axis of the nucleus. If this precession takes place rapidly compared with nuclear rotation (a point to be considered in more detail in Sec. III), the component of the nucleon angular momentum along the symmetry axis will be a constant of the motion. In analogy with the case of diatomic molecules, we designate this quantity by Ω which may take the values $j, j-1, \cdots, \frac{1}{2}$, where j is the magnitude of the single particle angular momentum.

Next, we must consider the motion of the nuclear core, its vibration and rotation. For our purpose the rotation is of particular significance. Even in the ground state, the nuclear axis will not stay fixed in space, but will undergo a precession around the total nuclear angular momentum vector. This motion, which is like that of a symmetrical top, may be characterized by the quantum numbers I, M, and K, representing, respectively, the magnitude of the total angular momentum, its projection on a fixed axis in space, and its projection on the nuclear symmetry axis. For the ground state, it is to be expected that $I=K=\Omega$ in which case the nuclear rotation energy is smallest.

The value of Ω corresponding to the lowest single particle state will depend on the couplings involved; but at any rate, for nuclei with one particle more or one particle less than required for a closed shell, the value $\Omega=j$ seems probable. In fact, in the two cases mentioned, the signs of the quadrupole moments show that the nuclear shape is of the type of an oblate and prolate spheroid respectively. In both cases therefore, the state of maximum Ω will be energetically favored.

Assuming $\Omega=j$, we have $I=j$ for the ground state. Thus the modification of the single particle model here considered need not impair the success of the model in predicting nuclear spins.

[4] E. Teller and J. A. Wheeler, Phys. Rev. 53, 778 (1938).

III. COUPLING OF ANGULAR MOMENTA AND DETERMINATION OF MAGNETIC MOMENTS

In a more detailed classification of nuclear states, which is required for the evaluation of magnetic moments in particular, it is necessary to compare the strengths of the couplings between the various angular momentum vectors of the nucleus.

In the first place, the coupling between the orbital and spin angular momenta, **l** and **s**, of the single particle, can be characterized by the frequency ν_{ls} with which the two vectors precess around each other. In the second place, the coupling of **l** to the nuclear axis may be described by the precession frequency ν_{lA} of **l** around the axis. Finally, it will be convenient to introduce the frequency ν_{AI} with which the axis precesses around **I**.

We consider first the case $\nu_{AI}>\nu_{lA}$, in which the coupling of the nucleon motion to the nuclear axis is weak. In this case, the single particle angular momentum and the rotation **R** of the nuclear core must first be quantized separately. For the ground state, R must be assumed to be zero and thus we have again essentially the extreme single particle model which does not allow for the existence of large quadrupole moments. We shall therefore assume $\nu_{AI}<\nu_{lA}$, which seems justifiable since, for the nuclear deformations in question, one expects values for $h\nu_{lA}$ of the order[3] of one Mev.

For $\nu_{AI}<\nu_{lA}$, a number of different cases arise, depending on the magnitude of the spin-orbit coupling:

(A). If the spin-orbit coupling is weak ($\nu_{ls}<\nu_{AI}$), the vector **l**, but not **s**, is coupled to the nuclear axis. The resultant of **l** and **R** forms a total orbital angular momentum **L** which is to be compounded with **s** by simple vector addition. For the lowest nuclear rotational state, the total orbital g-factor g_L will deviate only slightly from the single-particle value g_l, and the magnetic moment of the nucleus will therefore be close to the single particle value.

(B). For $\nu_{ls}>\nu_{AI}$, the total angular momentum of the single particle is coupled to the nuclear axis along which it has the component Ω. Assuming $\Omega=I$, one finds for the resultant nuclear g-factor:

$$g_I=g_\Omega\frac{I}{I+1}+g_R\frac{1}{I+1}, \tag{1}$$

where g_Ω and g_R are the g-factors associated with the angular momenta Ω and R. For g_R it seems most plausible to assume a value of about Z/A, although deviations from this value may occur, since only a fraction of the nucleons is expected to take part in the nuclear rotation. The value of g_Ω will depend on the relative magnitude of ν_{ls} and ν_{lA}.

(B_1). For $\nu_{ls}<\nu_{lA}$, the vector **l** is first coupled to the nuclear axis and, due to the spin-orbit coupling, the spin moment is subsequently quantized along the axis with a component $\pm\frac{1}{2}$. Consequently, we have:

$$Ig_\Omega = \pm\tfrac{1}{2}g_s + (I \mp \tfrac{1}{2})g_l \tag{2}$$

corresponding to the two possible types of states for a given I.

(B_2). If $\nu_{ls} > \nu_{lA}$, the vectors **l** and **s** must first be compounded to a resultant **j**, which is then coupled to the axis; in this case,

$$g_\Omega = g_j \tag{3}$$

where g_j is the Landé factor for l-s coupling.

In all three cases, A, B_1 and B_2, the model allows for the existence of large quadrupole moments. However, it should be noted that the quadrupole moment Q referred to fixed axes in space, for the substate $M=I$, will be smaller than the intrinsic nuclear quadrupole moment Q_0, defined with respect to the axis of the nucleus.

In the case A one thus finds:

$$Q = Q_0 \frac{I-1}{I+1}\,\frac{2I-1}{2I+1} \quad (I = L+\tfrac{1}{2}) \tag{4}$$

and

$$Q = Q_0 \frac{I}{I+1}\,\frac{2I-1}{2I+3} \quad (I = L-\tfrac{1}{2}) \tag{5}$$

whereas the cases B always lead to expression (5).

For small values of I, the ratio Q/Q_0 may be quite small; for $I=3/2$, the expressions (4) and (5) give values of 1/10 and 1/5 respectively.

The formulas given in the present section can be derived immediately from the vector model for compounding angular momenta, and from simple considerations regarding the motion of the symmetrical top. Wave functions corresponding to the different coupling cases can also be constructed, as regards their angular and spin dependence, in complete analogy to molecular wave functions.

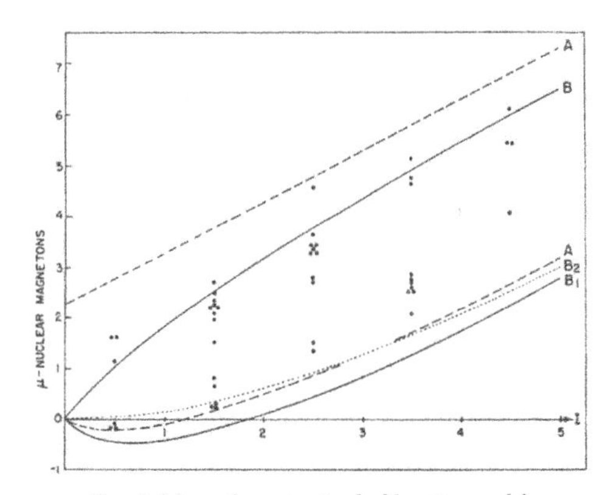

FIG. 1. Magnetic moments of odd proton nuclei.

IV. INTERMEDIATE COUPLING CASES

In Figs. 1 and 2, the magnetic moments of the odd nuclei are given.[5] Only nuclei with mass-number greater than 20 are included, since the models here discussed can only be assumed to have relevance for fairly heavy nuclei. The curves A represent the moments derived from the extreme single particle model (the Schmidt limits). As mentioned above, these values do not differ appreciably from those expected in coupling case A. The curves corresponding to the cases B_1 and B_2 are also shown in the figures, for a value of g_R of 0.45. For parallel **l** and **s** the expressions (2) and (3) lead to the same value for the magnetic moment.

It was one of the first outstanding successes of the single particle model that it accounted, in a general way, for the dependence of the moments on the value of I. Nevertheless, it is a major defect of the model that it fails to explain why the moments do not coincide with the values corresponding to pure single particle states, but lie in between the values for the two states possible for a given value of I.

This difficulty is especially conspicuous for nuclei having one particle more or one particle less than is required for a closed shell. In fact, the moments of these nuclei show no general tendency to approach the pure single particle values.

Since the two states in question have opposite parity, and therefore do not combine by means of conventional types of nuclear interaction, it is possible to alter the magnetic moments only by exciting additional particles. Such configuration perturbations would, however, have to be so strong that the single particle model would represent only a crude approximation. Moreover, it seems to be difficult to explain on this basis why the moments fall in between the two single particle values, rather than deviate more irregularly from these values.

The situation is essentially modified by the introduction of an asymmetric nuclear core. A large number of states now become possible for a given value of I. Thus, in the coupling case B the particular states considered in Sec. III were characterized by $\Omega = I$, and by the largest possible component of the single particle angular momentum along the nuclear axis. If these two restraining conditions are not imposed, it is possible to construct states having a wide range of moment values.

These values lie between definite limits. In fact, for a given value of the total orbital g-factor, g_L, the moment is uniquely determined by the component of the intrinsic spin along the axis of space-quantization. For the substate, $M=I$, the average value of this spin component S_Z satisfies the condition $-\frac{1}{2}I/(I+1) < \bar{S}_Z < +\frac{1}{2}$. Since the difference between g_L and g_l is usually

[5] The data are taken from a compilation of moments by H. L. Poss, published by Brookhaven National Laboratory, October 1949, supplemented by a few more recent moment determinations published in Phys. Rev.

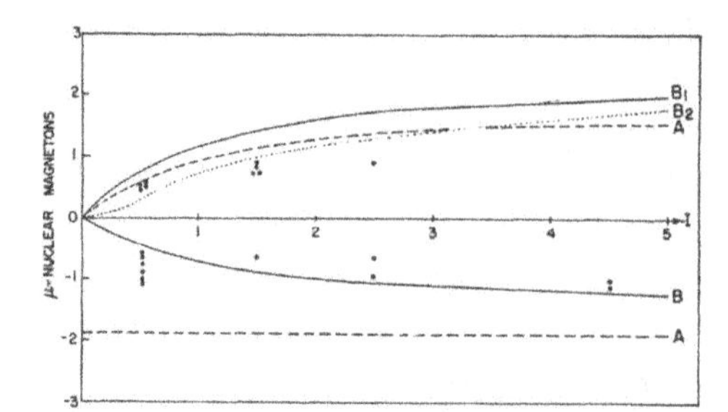

FIG. 2. Magnetic moments of odd neutron nuclei.

of minor importance, the moment is therefore confined within the limiting values A.

If the nucleus can be represented to a high approximation by one of the simple coupling cases discussed in Sec. III, the ground state would seem to have the quantum numbers which were assumed in that section, and which lead to the moment values given by the curves in Figs. 1 and 2. Not only is this choice of quantum numbers suggested by energetic considerations, but it is also required to make the nuclear spin equal to the single particle angular momentum.

However, when some of the characteristic frequencies introduced in Sec. III are comparable in magnitude, we have to do with mixed coupling types. The nuclear ground state will then no longer be a pure state in either of the representations A, B_1 or B_2, but will instead correspond to a mixture of states. Consequently the magnetic moments will, in general, fall in between the limiting values.

As an illustration, we may consider the particularly simple nuclei which have a single particle in addition to the number required for closed shells; e.g., ${}_{83}Bi^{209}$, and ${}_{51}Sb^{123}$. According to the shell model, which assumes that a strong spin-orbit coupling is a determining factor for the order of nuclear levels, these nuclei have $j=l-\frac{1}{2}$ and should be represented by the coupling case B_2. Nevertheless, the moments are found to lie very far from the predicted pure state values.

The influence of an asymmetric core implies, however, a coupling case intermediate between B_2 and B_1. The single particle state in question, having $\Omega=l-\frac{1}{2}$, will no longer be a pure $j=l-\frac{1}{2}$ state, but will have an admixture of $j=l+\frac{1}{2}$. The negative sign of the quadrupole moment, for the nuclei in question, indicates that the asymmetry favors states with large components of $\mathbf{l}$ along the nuclear axis. In this case, the magnetic moment for the state in question increases with increasing ν_{lA}. For $\nu_{lA}\gg\nu_{ls}$, the moment actually approaches the opposite limit B, corresponding to a spin-component of $+\frac{1}{2}$ along the nuclear axis. In order to account for the observed moments, comparable values of ν_{lA} and ν_{ls} are required. This appears to be consistent with the evidence[1] for a spin-orbit coupling of the order of 2 Mev for large values of l.

A detailed estimate of such perturbations is made difficult by the fact that a large asymmetry of the core will affect the ordering of the single particle levels to such an extent that the nuclei considered can no longer simply be characterized as having a single particle in addition to closed shells. Of course, this circumstance need not interfere with the explanation of the particular stability of the closed shell nuclei themselves, since these nuclei may be of spherical form.

If the deviation of the moments from the limiting values is ascribed to perturbations of the type considered, one would expect that nuclei having a j-value larger than that of any neighboring single particle level will have moments close to the pure state values. The known magnetic moments of nuclei of this type are given in Table I. There actually appears to be a rather close agreement with the values μ_B, predicted for the coupling case B. For comparison, the moments μ_A, corresponding to the extreme single particle model, are also listed. This model is not in good agreement with the empirical moments of these nuclei.

As regards the deviations of the moments from the μ_B-values, it must be noted that the moments depend to some extent on the value of g_R. As already mentioned, this value may vary from nucleus to nucleus, and a certain spread of the moments is to be expected. Moreover, even the simple nuclei considered will not represent completely pure states. In particular, the asymmetry of the core implies that l is not an exact quantum number. States of higher l, but with the same value of Ω, will therefore be admixed to some extent.

As an additional indication of the influence of the nuclear asymmetry on the magnetic moments, it may be noted that the following empirical rule seems to hold: of two isotopes, which differ by 2 neutrons and have the same value of I, the nucleus with the numerically smallest quadrupole moment has a magnetic moment closest to the pure state value.[6] Examples are

TABLE I. Magnetic moments of selected nuclei.

Nucleus	I	μ_{obs}	μ_B	μ_A
${}_{12}Mg^{25}$	5/2	−0.96	−1.04	−1.91
${}_{13}Al^{27}$	5/2	3.64	3.74	4.79
${}_{21}Sc^{45}$	7/2	4.75	4.86	5.79
${}_{23}V^{51}$	7/2	5.14	4.86	5.79
${}_{27}Co^{59}$	7/2	4.64	4.86	5.79
${}_{36}Kr^{83}$	9/2	−0.97	−1.19	−1.91
${}_{38}Sr^{87}$	9/2	−1.1	−1.19	−1.91
${}_{41}Cb^{93}$	9/2	6.17	5.92	6.79
${}_{49}In^{113}$	9/2	5.46	5.92	6.79
${}_{49}In^{115}$	9/2	5.47	5.92	6.79

[6] This rule is somewhat different from the one suggested by W. Gordy (Phys. Rev. **76**, 139 (1949)). In many cases the two rules coincide but, although the evidence is not conclusive, Gordy's rule appears not to hold for the Cu-isotopes (P. Brix, Zeits. f. Physik **126**, 725 (1949)).

the isotopes of Cl, Cu, Ga, Br, Eu, and Re. The only known exception is In, whose isotopes, however, have almost identical moments.

V. CONCLUDING REMARKS

The most adequate coupling cases to represent nuclear states may be expected to vary from nucleus to nucleus depending, in particular, on the quadrupole moment and the value of the single particle angular momentum.

As regards the question of predominant coupling types, some information can be obtained from Figs. 1 and 2.

In the first instance, a large part of the moments for small values of I fall outside the curves B_2. Apart from the case $I=\frac{1}{2}$ with parallel $\mathbf{l}$ and $\mathbf{s}$, it appears, however, that practically all moments of heavy nuclei lie within the limits B_1, although these limits are considerably narrower than the A curves. In view of the uncertainty of g_R, the only definite exception is the moment of Pr^{141} ($I=5/2$), which lies very close to the A-limit.

It is not surprising that the moments for $I=\frac{1}{2}$ with parallel $\mathbf{l}$ and $\mathbf{s}$ tend to fall outside the B curves. In fact, just in this particular case, the projection of $\mathbf{l}$ on the nuclear axis vanishes (Σ states in molecular terminology). There is thus no net spin-orbit coupling, and the coupling of $\mathbf{s}$ to an axis of the nucleus is presumably very weak. Consequently, an approach to the model A is to be expected.

Evidence regarding the structure of nuclear magnetic moments has been obtained from a study of the anomalies in atomic hyperfine structures caused by the finite size of the nucleus.[7,8] The analysis of such effects has given indications that the orbital g-factor of the nucleus is close to the single particle value. This evidence is in agreement with all of the models considered here, since for the lowest rotational state of the nucleus the total orbital g-factor, g_L, deviates only slightly from g_l. It may be added that this value for g_L would not be obtained in general if the large nuclear quadrupole moments were ascribed to more complicated departures from the single particle model than those discussed here.

The hfs anomaly for the odd K-isotopes[8] allows a more detailed test of the nuclear model. These nuclei contain 19 protons, one less than the number 20 required for closed shells, and are therefore expected to be well represented by the model of a single particle moving in an asymmetric nuclear core. Assuming a coupling case between B_1 and B_2, the nuclear wave functions can be determined from the empirical values of the moments. An estimate of the hfs anomaly on this basis[9] gives a result in close agreement with the observed value.

In many respects it thus appears that the modifications of the extreme single particle model, which are necessary in order to allow for the large quadrupole moments, at the same time improve the agreement of the model with empirical data on nuclear magnetic moments.

The writer is indebted to Columbia University for a fellowship enabling him to take part in research work in the Physics Department, and wishes to acknowledge the benefit of stimulating discussions with Professors I. I. Rabi, J. L. Rainwater, and C. H. Townes on the topics of the present paper.

[7] A. Bohr and V. Weisskopf, Phys. Rev. **77**, 94 (1950).
[8] Ochs, Logan, and Kusch, Phys. Rev. **78**, 184 (1950).
[9] A. Bohr, Phys. Rev. (to be published).

Aage and Niels Bohr in Copenhagen in 1954, after Aage's thesis defence. Courtesy of Niels Bohr Archive, Copenhagen.

ROTATIONAL STATES OF ATOMIC NUCLEI

By

AAGE BOHR

Universitetets institut for teoretisk fysik

KØBENHAVN
EJNAR MUNKSGAARDS FORLAG
1954

This paper, which gives a survey of the theory of nuclear rotational states and of the experimental evidence concerning this type of nuclear excitation, is based upon studies begun during the author's stay at Columbia University in 1949—50, and since then continued at the Institute for Theoretical Physics of the University of Copenhagen. I would like to express my thanks to the staff of the Physics Department of Columbia University, especially to professors Rabi, Rainwater, and Townes, as well as to members and guests of the Institute for Theoretical Physics, for many stimulating discussions. I am deeply indebted to my father, professor Niels Bohr, for what he has taught me of science and for the encouragement and support he has given me in my work. Many of the results here reported were obtained in co-operation with Dr. Ben R. Mottelson and I wish to give expression to the great stimulus and pleasure this co-operation has given me. I would also like to thank Mrs. S. Hellmann for her untiring help with manuscripts and proofs of this, as well as of previous publications.

Aage Bohr

Copenhagen, March 1954.

Contents.

I. Introduction. Atoms and Nuclei.

In the atomic nucleus, we are dealing with a system of particles, the nucleons, held together by their mutual attractive forces. In the attempt to analyze the structure of such a quantum mechanical many-particle system, it is instructive to draw comparisons with the more familiar problem of the electronic constitution of atoms. Although the cohesive forces in the two systems are of essentially different character, important similarities in the structure of nuclei and atoms arise from the fact that, in both cases, we have to do with systems of particles whose behaviour is governed by the exclusion principle.

It is well known that a simple understanding of many properties of atoms is obtained by considering the electrons as moving approximately independently in an average atomic potential. Each binding state in this potential can be occupied by only a single electron, and the ground level of the atom is obtained by filling the electronic states of lowest energy. States having approximately the same energy, such as those differing only in their angular orientation, are said to form a "shell", and when all the states of a shell are filled, one talks of a closed shell of particles. Many properties of the atom, such as the ionization energy, the valency, etc., are intimately correlated with the shell structure and depend primarily on the number of particles outside of closed shells. Thus, the atoms whose electrons all form closed shells possess special stability and are chemically inert; the noble gases have just this type of structure.

A similar order appears to govern the motion of the protons and the neutrons in the nucleus. Especially in recent years, numerous regularities in the nuclear properties have been found, which are indicative of a shell structure. In particular, nuclei possessing certain numbers of neutrons or of protons are found to exhibit a pronounced stability, and are interpreted as representing the closed-shell configurations. A nuclear shell model has been developed, which considers the nucleons as moving approximately independently in a nuclear binding field and, for suitable assumptions about this field, it has been possible to account successfully for many of the observed regularities in nuclear properties (MAYER 1950; HAXEL, JENSEN, and SUESS, 1950, 1952).

The approximate validity of the atomic and nuclear shell model is to be understood from the fact that the interactions between the particles may be regarded as contributing only to an averaged binding field, and do not appreciably perturb the motion of the individual particles in this field. In the usual formulation of the shell model, the binding field is treated as a fixed quantity, characteristic of the given state of the system; but, to the extent that the field is produced by the particles themselves, it must actually be recognized as a dynamical variable. Variations in the field are associated with a collective motion of the particles, and one is thus led to consider a generalization of the shell model, which incorporates collective oscillations of the system as a whole.

Indeed, already at an early stage of nuclear theory, it was pointed out that a system in which the particles are held together by their mutual interactions must always be expected to exhibit modes of motion involving the particles collectively. Apart from the center of gravity motion, the shape and size of such a system, and its density distribution, may oscillate in a manner resembling the collective modes of matter in bulk (N. BOHR 1936; N. BOHR and F. KALCKAR, 1937).

In the atomic case, the dynamic aspects of the field are of relatively minor significance, due to the stabilizing influence of the central nucleus. The attraction from the nucleus plays a dominant role in shaping the atomic field and, due to the large mass of the nucleus as compared with that of the electrons, the nuclear position is very nearly fixed. Moreover, the atomic shell model has found its most detailed applications to the X-ray and optical spectra. In the former case, the field is almost entirely due to the nucleus and, in the latter case, the loosely bound outer electrons can affect only slightly the charge distribution in the tightly bound atomic core[1].

In the nucleus, however, there is no similar stabilizing agent, and variations in the field associated with collective modes of motion play an essential role. Thus, the nucleus must be considered as a deformable system (RAINWATER 1950), and its shape and orientation are to be regarded as dynamical variables (A. BOHR 1951). The motion of the individual particles is strongly affected by the collective oscillations, and the nuclear properties are determined by the interplay of these two basic types of motion. In some respects, the dynamics of the nucleus is analogous to that of molecules where, similarly, we have to do with the interplay between the motion of the individual electrons, and vibrations and rotations of the structure as a whole (A. BOHR 1951, 1952).

An attempt has been made to develop in a consistent way such a unified collective and individual-particle description of the nucleus[2,3,4]. The analysis of the empirical data has provided evidence that the collective aspects of nuclear dynamics play an important part in a large variety of nuclear phenomena.

This is, perhaps, most strikingly illustrated by the nuclear excitation spectrum, where the degree of freedom associated with the collective motion implies new types of stationary states which have no analogue in the atomic spectrum. In particular, nuclei with large deformations possess a very simple type of motion, which merely affects the orientation of the axes of the nuclear shape, and which gives rise to a spectrum, similar in various respects to that of a rotating rigid body. Within the last year, an extensive empirical material has become available, which attests to the remarkable features of these nuclear states.

The present paper gives an outline of the theory of nuclear rotational states and a discussion of the empirical evidence. Especially, it will be attempted to give a simple description of the main physical qualities of the rotational spectrum. For mathe-

[1] The significance of collective electronic motion for stopping power problems, which involve the mean atomic excitation energy, has recently been discussed by LINDHARD and SCHARFF (1953). Cf. also BLOCH (1933).

[2] A. BOHR (1952); in the following to be referred to as A, its sections and equations as (A, § V. 2), (A, 37), etc.

[3] A. BOHR and B. R. MOTTELSON (1953); to be referred to as BM.

[4] A unified description of nuclear structure has also been given by HILL and WHEELER (1953).

matical details, the reader is referred to the previous publications, where the general theory of collective nuclear motion is developed.

The existence of rotational spectra is intimately associated with the large nuclear deformations, and we therefore begin, in Section II, with a brief discussion of nuclear deformations. Section III describes the nuclear rotational motion, which differs fundamentally from a rigid rotation and can best be characterized as a wave travelling around the nucleus. The energy spectrum is considered in Section IV and compared with available empirical data. Section V deals with electromagnetic transitions, the study of which provides an important tool in the identification of rotational states. Section VI contains a few concluding remarks.

In the Appendix, we discuss the derivation of the hydrodynamical equations for the collective flow in a rotating nucleus.

II. Nuclear Deformations.

It has been known for a long time that many nuclei possess very large deformations. This was first discovered in the study of the hyperfine structure of atomic spectra (cf., e. g., KOPFERMANN 1940). For a non-isotropic nuclear charge distribution, the energy of the atom depends on the orientation of the electronic orbits with respect to the nucleus and, although the effect is very small, it has been detected for a large number of nuclei. More recently, the phenomenon has also been observed in the spectra of molecules and crystals.

The effect determines the nuclear quadrupole moment Q defined by

$$eQ = \langle \int \varrho_e (3 z^2 - r^2) \, d\tau \rangle_{M = I}, \tag{1}$$

where ϱ_e is the nuclear charge distribution, and e the charge unit. The moment is evaluated for a state of orientation of the nucleus with magnetic quantum number M equal to the nuclear spin I. The quadrupole moment may be said to characterize the degree of spheroidal anisotropy which the nuclear state possesses.

The available empirical evidence on quadrupole moments in the region of the heavy elements is shown in Fig. 1. Here, as in the following, we restrict ourselves to nuclei with mass number $A > 140$, which provide the main evidence on the rotational spectrum.

The figure illustrates two significant features of the quadrupole moments. One is their magnitude; for most of the nuclei in the figure, the moment is more than ten times larger than the contribution of any single proton bound in the nuclear field. We are clearly observing a co-operative effect of a very large number

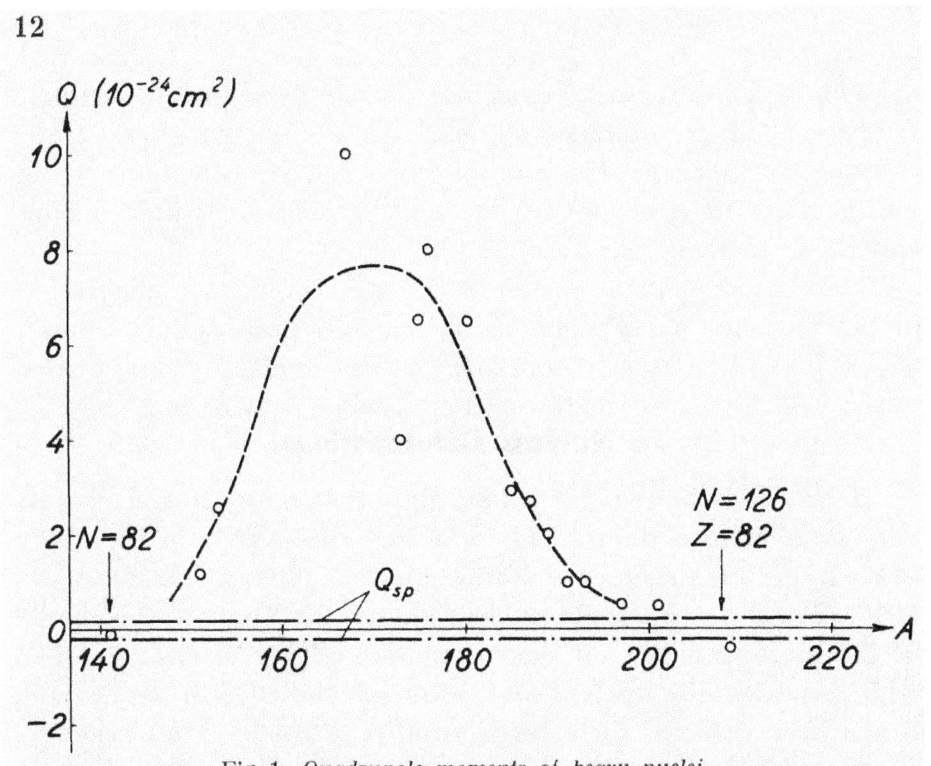

Fig. 1. *Quadrupole moments of heavy nuclei.*
The empirical quadrupole moments Q are plotted as a function of the nuclear mass number A for $A > 140$. The broken, approximately horizontal lines indicate the magnitude of the single-particle moment Q_{sp} to be expected if the nuclear angular momentum were concentrated on a single proton. The magnitude of Q_{sp} depends somewhat on the nuclear spin, I, and the estimate refers to $I = 5/2$, which is the mean value of the observed spins; the sign of Q_{sp} is negative for a single particle outside of closed shells, positive for a particle lacking in a closed shell. Arrows indicate the doubly closed-shell configurations of $_{82}Pb^{208}$ ($Z = 82$ and $N = 126$) and the closed neutron configuration of $_{59}Pr^{141}$ ($N = 82$), in the regions of which the quadrupole moments are small. The broken curve indicates the average trend of Q between the closed-shell regions. For references to the Q-measurements and for estimates of Q_{sp}, cf. BM (Chapter V and Addendum to Chapters IV and V). For the recent measurement of Q (Pr^{141}), cf. Lew (1953).

of particles, and it is appropriate to talk about a deformation of the nucleus as a whole.

The other characteristic feature of the quadrupole moments is their systematic variation with A. They are largest for $A \approx 170$ and decrease as we move away from this region, reaching small values for $A \approx 210$ and also for $A \approx 140$. This variation is associated with the nuclear shell structure; for the nucleus $_{82}Pb^{208}$, both protons and neutrons form closed shells, and again for nuclei with A near 140, the neutrons form closed shells. Thus, generally speaking, the nuclear deformations are the larger, the further we are removed from closed-shell structures.

The magnitude as well as the variation of the quadrupole moments can be understood as a simple consequence of the independent-particle structure of the nucleus, when one takes into account the freedom of the nuclear field to assume a non-spherical shape.

The preferred shape of the field depends on the configuration of the particles. Only if the neutrons and protons form closed shells, the self-consistent treatment of the nucleus yields an isotropic field. In this case, the particles, when moving in a spherically symmetric potential, possess themselves a spherically symmetric density distribution.

If particles are added to, or subtracted from, a closed-shell configuration, the particle density is no longer isotropic. The extra particles tend to deform the nucleus so as to make the attractive field conform better to their own density distribution, and the shape of the nucleus depends on the balance between the deforming power of the extra particles and the restoring force of the nuclear core (Rainwater 1950). The effect may also be illustrated in terms of the centrifugal pressure which a bound particle exerts on the nuclear wall, in the plane of the orbit.

This polarization of the nucleus may lead to a quadrupole moment many times larger than that contributed by the deforming particles themselves. Moreover, the particles act co-operatively, since they all prefer to orient their orbits so as to conform as well as possible with the nuclear shape and, therefore, exert their pressure on the wall in the same direction. Thus, the nuclear deformation increases rapidly as we move away from closed shells. As the next closed shell is approached, the only available orbits are oriented in such a way as to counteract the deformative pressure, and the nuclear deformation decreases again.

While it is possible in this way to account for the qualitative features of nuclear quadrupole moments, no detailed application of the self-consistent treatment extended to non-spherical fields has so far been made. Even the more restricted problem of the radial density distribution in a spherical nucleus, and its dependence on the nuclear forces, presents many difficulties and has not yet been explored very far.

Estimates of nuclear quadrupole moments have been made on the basis of a simplified model, assuming the nucleus to have a constant density, and the nuclear deformability to be a smooth function of A, which can be derived from the empirical nuclear binding energies[1]. These estimates account for many trends in the quadrupole moments and also for their general order of magnitude, although they somewhat exceed the empirical values (cf. van WAGENINGEN and de BOER, 1952; van WAGENINGEN 1953, and BM, Chapter V).

Such considerations also show that most nuclei prefer a shape of cylindrical symmetry. Only for very special configurations it seems that an asymmetric shape may be more favourable (cf. BM, § II c.ii; B. SEGALL, private communication).

For the interpretation of quadrupole moments, it is of significance that the shape of the nucleus is not directly related to the spectroscopically determined quantity (1), but is characterized by the "intrinsic" quadrupole moment Q_0, defined by

$$eQ_0 = \int \varrho_e (3 z'^2 - r'^2) \, d\tau , \tag{2}$$

where the primed co-ordinates refer to a co-ordinate system fixed in the nucleus; the nuclear symmetry axis is chosen as z'-axis. Thus, for a uniformly charged nucleus of spheroidal shape,

$$Q_0 = \frac{4}{5} Z R_0^2 \cdot \frac{\Delta R}{R_0}, \tag{3}$$

where Z and R_0 are the charge number and average nuclear radius, while ΔR is the difference between the major and minor semi-axes. In expression (3) it is assumed that $\Delta R \ll R_0$.

In a quantum state with given I and M, the directions of the nuclear axes have a certain indeterminacy, which tends to diminish the effective charge asymmetry of the state. Therefore, the magnitude of Q, given by (1), is somewhat smaller than that of Q_0, given by (2); the quantitative relationship will be discussed below (§ V. a).

From the present interpretation of the nuclear quadrupole moments, one would expect a rather smooth variation of the nuclear deformation with A, more so than the data in Fig. 1 might indicate. A plot of Q_0 rather than Q only shows slightly smaller fluctuations. One must, however, bear in mind the considerable uncertainty which attaches to most experimental quadrupole determinations, and the magnitude of which is difficult to estimate.

[1] A somewhat different model has been used by PFIRSCH (1952) in his study of the trends of the quadrupole moments.

III. The Collective Rotational Motion.

A deformed system possesses an especially simple type of motion which merely consists of a rotation of the orientation in space. If performed sufficiently slowly, this motion will not affect the internal structure of the system. The energy E_{rot} associated with the rotation is then purely kinetic and is proportional to the square of the angular velocity ω, so that we may write

$$E_{\mathrm{rot}} = \frac{1}{2} \mathfrak{J} \, \omega^2 . \tag{4}$$

While the form of this expression is very general, the parameter $\mathfrak{J}$, which may be termed the effective moment of inertia, depends on the structure of the system.

In the case of an approximately rigid structure, in which the particles are confined to small regions around their equilibrium positions, each particle must partake in the rotation with the angular velocity ω. The moment of inertia is then given by the familiar expression

$$\mathfrak{J}_0 = \sum_p m_p r_p^2 , \tag{5}$$

where m_p and r_p denote the mass of the pth particle and its distance from the rotational axis.

If the structure is not rigid, the particles have greater freedom in their motion and the system may adjust more easily to the changing orientation. The rotation is then accomplished by smaller displacements of the particles, and the effective moment of inertia is smaller than (5).

The structure of a nucleus may be said to be the extreme counterpart to that of a rigid body, since the nucleons freely traverse the entire nuclear volume. For such a system, the rotational motion may be characterized as a wave travelling around the nucleus; this type of motion accomplishes the rotation with the minimum kinetic energy.

The wave-like rotational motion can be described in terms of a collective flow $\vec{v}(\vec{x})$ governed by the hydrodynamical equations

$$\text{rot}\,\vec{v} = 0 \qquad \text{div}\,\varrho\vec{v} = -\frac{\partial\varrho}{\partial t} = \omega\frac{\partial\varrho}{\partial\varphi}, \tag{6}$$

where $\varrho(\vec{x})$ is the density of particles, and where φ is the azimuth with respect to the axis of rotation. The irrotational character of the nuclear velocity field is a consequence of the long mean free path of the particles characteristic of a shell structure. The introduction of a collective flow in a quantum mechanical many-particle system and the derivation of the hydrodynamical equations (6) are discussed in the Appendix.

The difference between the velocity fields associated with the rigid and wave-like rotations is illustrated in Fig. 2.

A classical analogue to the nuclear rotational motion is afforded by a surface wave in a liquid drop, even though the two forms of matter have an essentially different structure. However, we are not concerned with a free surface wave on a spherical surface, which would be characterized by a definite proper frequency. Rather, we must consider a drop which, because of intrinsic strains, possesses a non-spherical equilibrium shape. According to classical hydrodynamics, the geometrical shape of such a body can be set in rotation with an arbitrary frequency.

For the wave-like rotation, the collective velocity at each point is proportional to the amplitude of the wave, i.e., to the deformation of the system. The effective moment of inertia of the nucleus is thus proportional to the square of the deformation, and we may write

$$\mathfrak{J} \approx \mathfrak{J}_0\,\beta^2, \tag{7}$$

where β is a dimensionless parameter characterizing the deformation with respect to the axis of rotation.

For a known nuclear density distribution, a quantitative determination of $\mathfrak{J}$ can be obtained by solving the hydrodynamical equations (6) and using the relation

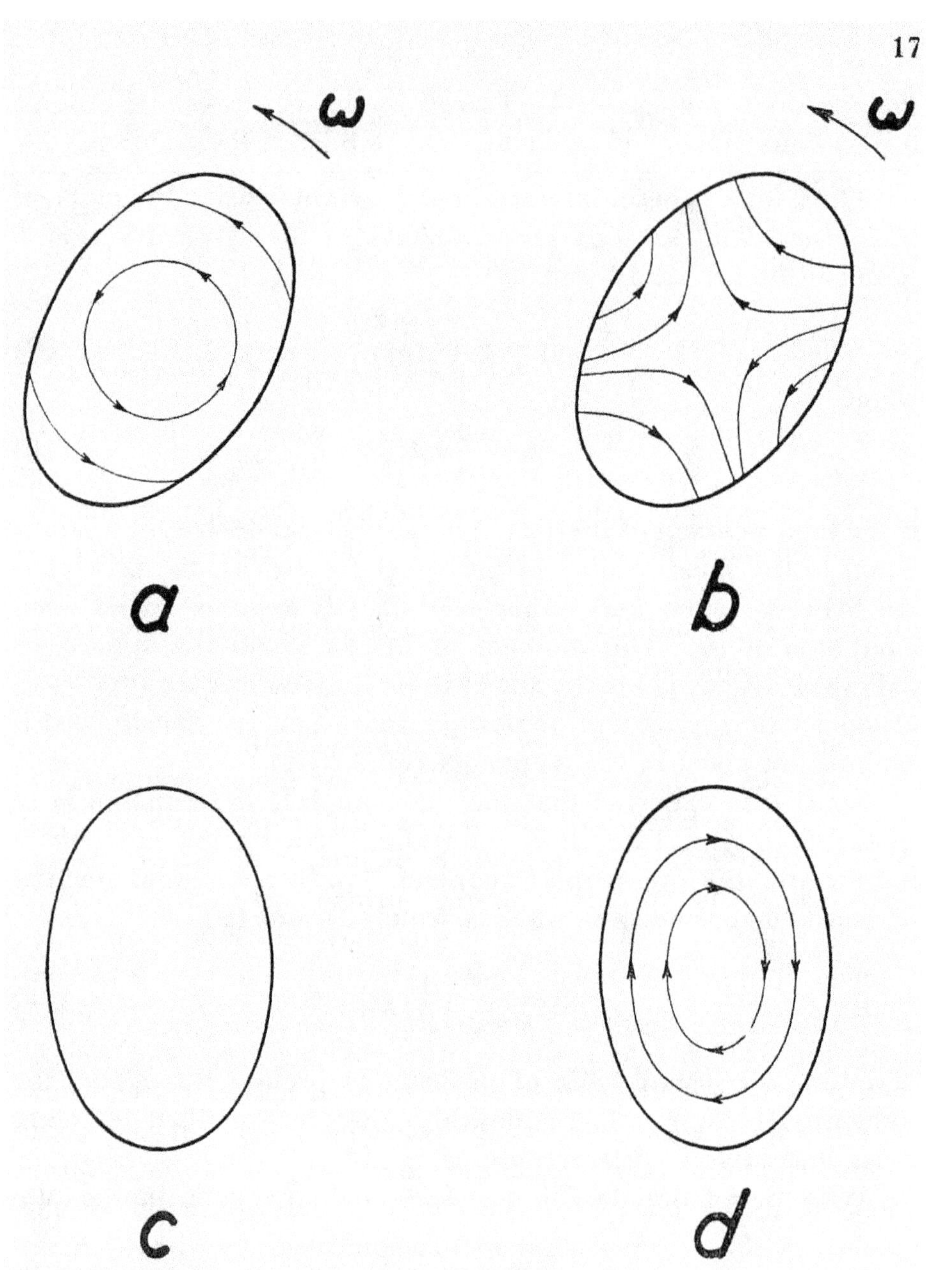

Fig. 2. *Velocity fields for rigid and wave-like rotations.*
The two above figures illustrate the streamline picture, seen from a fixed co-ordinate system. Fig. 2a refers to the rigid rotation, in which the particles move in circles around the axis of rotation. In the wave-like rotation (b) the particles perform oscillatory motions and only the geometrical shape rotates.
The two figures below show the same motions seen from the rotating co-ordinate system, in which the density distribution remains constant. For the rigid rotation (c) the velocities vanish, while for the wave-like rotation (d) the streamlines form closed loops and the particles circulate in the direction opposite to the rotation.

$$E_{\text{rot}} = \frac{1}{2} m \int \varrho \vec{v}^2 \, d\tau = \frac{1}{2} \mathfrak{J} \omega^2. \tag{8}$$

Thus, for a spheroidal nucleus of constant density, the moment of inertia about an axis perpendicular to the symmetry axis is found to be (cf. A, 16 and 28)

$$\mathfrak{J} = \mathfrak{J}_0 \left(\frac{\Delta R}{R_0} \right)^2, \tag{9}$$

where

$$\mathfrak{J}_0 = \frac{2}{5} m A R_0^2 \tag{10}$$

is the total moment of inertia (5) for a spherical nucleus of a radius equal to the mean radius of the spheroid. In (9), the difference ΔR between major and minor semi-axes is assumed to be small compared to R_0. The moment of inertia about the spheroidal axis vanishes, as is always the case for a symmetry axis (cf. (7)). Moments of inertia for somewhat more general density distributions are given in the Appendix (cf. XXIX).

Since it is expected that the main nuclear deformation is of quadrupole type, there is a connection between the moment of inertia and the quadrupole moment. For a spheroidal nucleus of constant density, one obtains from (3) and (9)

$$\mathfrak{J} = \frac{m^2}{4 \mathfrak{J}_0} \left(\frac{A}{Z} \right)^2 Q_0^2, \tag{11}$$

assuming a constant ratio of neutrons to protons throughout the nucleus. However, this relationship may be modified for other mass and charge distributions.

With the rotational motion is associated an angular momentum

$$\mathfrak{P} = \mathfrak{J} \omega. \tag{12}$$

The existence of a net angular momentum for a flow of the type illustrated in Fig. 2b is connected with the fact that more nuclear matter is concentrated in the directions where the flow follows the rotation than in the directions where it goes against the rotation.

The expressions (4) and (12) refer to the rotation about a fixed axis. An arbitrary rotation of the nucleus may be conveniently resolved into its components along the principal inertial axes. Denoting the moments of inertia along these axes by $\mathfrak{J}_\varkappa$ ($\varkappa = 1, 2, 3$), and the components of the angular momentum by $\hbar R_\varkappa$, one obtains, just as for the rigid rotation,

$$E_{\text{rot}} = \sum_\varkappa \frac{\hbar^2 R_\varkappa^2}{2\,\mathfrak{J}_\varkappa}. \tag{13}$$

The rotational properties are greatly simplified if the nuclear deformation possesses axial symmetry, as is in general expected to be the case (cf. above, p. 14). Since the moment of inertia vanishes about the symmetry axis, the nucleus does not rotate collectively around this axis. The expression (13) then reduces to

$$E_{\text{rot}} = \frac{\hbar^2}{2\,\mathfrak{J}} (\vec{R})^2, \tag{14}$$

where $\mathfrak{J}$ is the moment of inertia about an arbitrary axis perpendicular to the symmetry axis.

IV. The Rotational Energy Spectrum.

a) Theoretical expressions.

In addition to the angular momentum of the collective rotation, the nucleus may possess an intrinsic angular momentum associated with the motion of individual particles.

The binding states of the nucleons in a deformed, but cylindrically symmetric field may be characterized by their component Ω of angular momentum along the symmetry axis. The quantum number Ω takes on half-integer values, positive and negative. The states are always two-fold degenerate, since a change of sign of Ω simply corresponds to a reversal of the sense of motion about the axis, which has no influence on the energy of the state.

In the case of a nucleus with an even number of neutrons and an even number of protons, the particles thus fill pairwise in states with opposite sign of Ω, and the net particle angular momentum vanishes. The total nuclear angular momentum $\vec{I}$ is then equal to that of the collective rotational motion, and since $(\vec{I})^2$ always has the proper values $I(I+1)$, one gets from (14) the simple expression

$$E_{\text{rot}} = \frac{\hbar^2}{2\mathfrak{J}} I(I+1) \tag{15}$$

for the rotational spectrum[1].

The wave function describing the rotation of the even-even nucleus may be expressed in terms of the polar angles (Θ, φ) of the nuclear axis and are given by

[1] Although the total particle angular momentum in an even-even nucleus vanishes on the average, there are fluctuations in its components perpendicular to the nuclear symmetry axis. This implies that $(\vec{R})^2$ actually differs somewhat from $(\vec{I})^2 = I(I+1)$ (cf., e. g., A, 98), but, since this difference is independent of I, it does not affect the rotational spectrum.

$$\Phi_{I,M} = Y_{I,M}(\Theta, \varphi), \tag{16}$$

where M is the magnetic quantum number and Y the normalized spherical harmonic. The angles Θ, φ are collective nuclear co-ordinates, which are to be regarded as functions of the nucleonic co-ordinates (cf. the Appendix). The total nuclear wave function is a product of (16) and a wave function describing the intrinsic structure of the nucleus (cf. (XI), p. 47).

The nuclear deformation is symmetric with respect to a reflection in the nuclear center, and this restricts the possible quantum states of rotation in a similar manner as for molecules with identical nuclei. Thus, a given set of co-ordinates of the nucleons only defines the nuclear orientation (Θ, φ) to within a rotation by 180°; it does not define a direction of the nuclear axis. Consequently, the wave function, being a one-valued function of the particle co-ordinates, must be invariant to such a rotation. The permissible values of I are, therefore,

$$I = 0, 2, 4, 6 \ldots . \tag{17}$$

The polar angles Θ and φ are even functions of the particle co-ordinates so that the wave function (16) always has even parity. Since the internal nuclear wave function, due to the filling of the states in pairs, also has even parity, the rotational states (17) all have even parity.

In an odd-A nucleus, the last odd particle contributes an angular momentum Ω along the nuclear axis. The coupling diagram for the rotational states of such a nucleus is shown in Fig. 3, from which one obtains $(\vec{R})^2 = (\vec{I})^2 - \Omega^2$. Thus, apart from a constant term, the rotational energy is again given by (15); denoting the ground state spin by I_0, the excitation energy becomes

$$E_{\text{rot}} = \frac{\hbar^2}{2\mathfrak{J}}(I(I+1) - I_0(I_0+1)). \tag{18}$$

The magnitude of I must now be at least equal to its component along the axis so that the ground state has $I_0 = \Omega$.

For the odd-A nucleus, the wave function has a similar structure as that of a diatomic molecule, in which the electrons contribute a net angular momentum along the molecular axis (cf.,

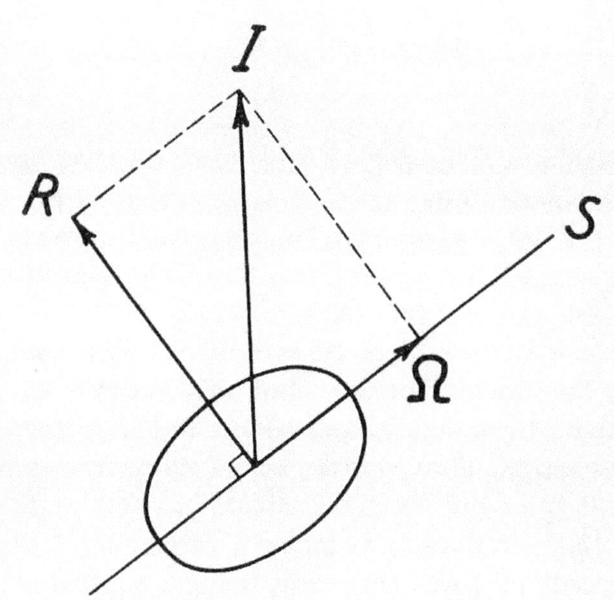

Fig. 3. *Coupling scheme in an odd-A nucleus.*
In a deformed nucleus of axial symmetry, the binding states of the individual particles are characterized by their components of angular momentum about the nuclear axis S. In an even-even nucleus, the particles fill pairwise in degenerate states, having opposite angular momenta along the axis, and so give no net intrinsic angular momentum to the nucleus. In an odd-A nucleus, the last odd particle contributes an intrinsic angular momentum Ω along the axis, and the total nuclear angular momentum $\vec{I}$ is the sum of this intrinsic angular momentum and the collective rotational angular momentum $\vec{R}$. The latter is perpendicular to S, since the nucleus cannot rotate collectively about a symmetry axis.

e. g., BM, II. 15). Such a system in general possesses two degenerate states for each I, corresponding to opposite signs for Ω; however, only for a certain combination of these two states, the wave function is invariant to an inversion of the direction of the nuclear axis (cf. A, § V. 2). Thus, the nucleus possesses only one rotational state for each I, and the sequence of states is

$$I = I_0, I_0 + 1, I_0 + 2, \ldots . \quad (19)$$

As in the even-even nucleus, the total nuclear parity is equal to that of the internal nuclear wave function. Thus, all levels in the rotational spectrum (19) have the same parity as the binding state of the last odd particle.

The symmetry of the nuclear shape may have further consequences if the last odd particle has $\Omega = 1/2$. The orbital angular momentum of the particle may then have a vanishing component along the nuclear axis, part of the time, in which case there are no forces to couple the particle intrinsic spin to this axis. The spin is then decoupled from the rotational motion. This effect may be taken into account by adding to the rotational energy (15) the extra term (BM, VI. 5)

$$E'_{\text{rot}} = (-)^{I+1/2} \frac{\hbar^2}{2\mathfrak{J}} \left(I + \frac{1}{2}\right) \sum_j (-)^{j-1/2} c_j^2 \left(j + \frac{1}{2}\right), \tag{20}$$

where c_j^2 represents the probability that the odd particle has a total angular momentum j.

The extra rotational energy (20) implies an anomalous rotational spectrum for this class of nuclei and may lead to a different sequence of states than given by (19). For example, if the particle is predominantly in a state with $j = 3/2$ (and $\Omega = 1/2$), the nuclear ground state becomes $I = 3/2$ with $I = 1/2$ and $I = 7/2$ as the two first excited states.

Finally, in an odd-odd nucleus, the last proton as well as the last neutron contribute angular momenta along the nuclear axis. The total intrinsic nuclear angular momentum may be the sum or difference of these two contributions, and one thus obtains two rather close-lying rotational families. Each has a spectrum given by (18) and (19).

As already emphasized, the simple expression (4) for the rotational energy, which leads to a spectrum of the type (15) or (18), represents a limiting situation realized only when the rotational motion is so slow that it does not appreciably perturb the intrinsic structure of the nucleus. With increasing rotational frequency, the particle structure can no longer follow adiabatically the changing orientation of the field; moreover, the nuclear deformation increases, due to a centrifugal effect similar to the vibration-rotation interaction in molecules.

These perturbations give rise to correction terms to the rotational spectrum. One of the most important of these is expected to arise from the increase, with I, of the deformation, and thus of the moment of inertia. This effect gives a negative contribution to E_{rot}, the magnitude of which depends on the nuclear deformability, and can be expressed in terms of the frequencies of vibration of the nucleus about its equilibrium shape. As regards orders of magnitude, one finds

$$\Delta E_{\text{rot}} \approx -2\left(\frac{1}{\hbar\omega_{\text{vib}}}\right)^2\left(\frac{\hbar^2}{\mathfrak{J}}\right)^3 I^2(I+1)^2, \tag{21}$$

where ω_{vib} is a measure of the vibrational frequencies. The more detailed estimate of the effect involves an analysis of the proper modes of vibration of the deformed nucleus (cf. BM, § VIc. ii).

The importance of correction terms of the type (21) depends on the ratio of rotational to vibrational energies, which again depends on the nuclear deformation. While $\hbar\omega_{\text{vib}}$ is approximately independent of the deformation, and is expected to be on the average about one or two MeV for a heavy nucleus, although somewhat dependent on the nucleonic configurations, the rotational energy quantum decreases with the square of the deformation (cf. (7)). The two energies are about equal when the deformation is comparable with the zero point amplitude of the vibrational oscillations.

Thus, the existence of a rotational spectrum is characteristic of a system whose equilibrium deformation is large compared with the fluctuations in the deformation. When this condition is not fulfilled, one cannot, for a non-rigid structure like the nucleus, separate a collective rotational motion from the internal degrees of freedom.

b) Empirical data.

We first consider the even-even nuclei, for which the rotational spectrum is especially easy to identify, due to its great regularity.

The sequence of states is given by (17) and is, in the first place, in accordance with the well-known empirical rule that all even-even nuclei have a ground state spin of zero. The spectrum (17) applies only to deformed nuclei, but an analysis of the coupling scheme for nuclei with approximately spherical equilibrium shapes also yields a ground state spin of zero, under very general conditions (cf. BM, § III.ii, which also contains references to previous investigations of this problem).

Moreover, it has been found in recent years that the first excited states in even-even nuclei have a spin of two and even parity (2 +), just corresponding to the rotational spectrum (cf.,

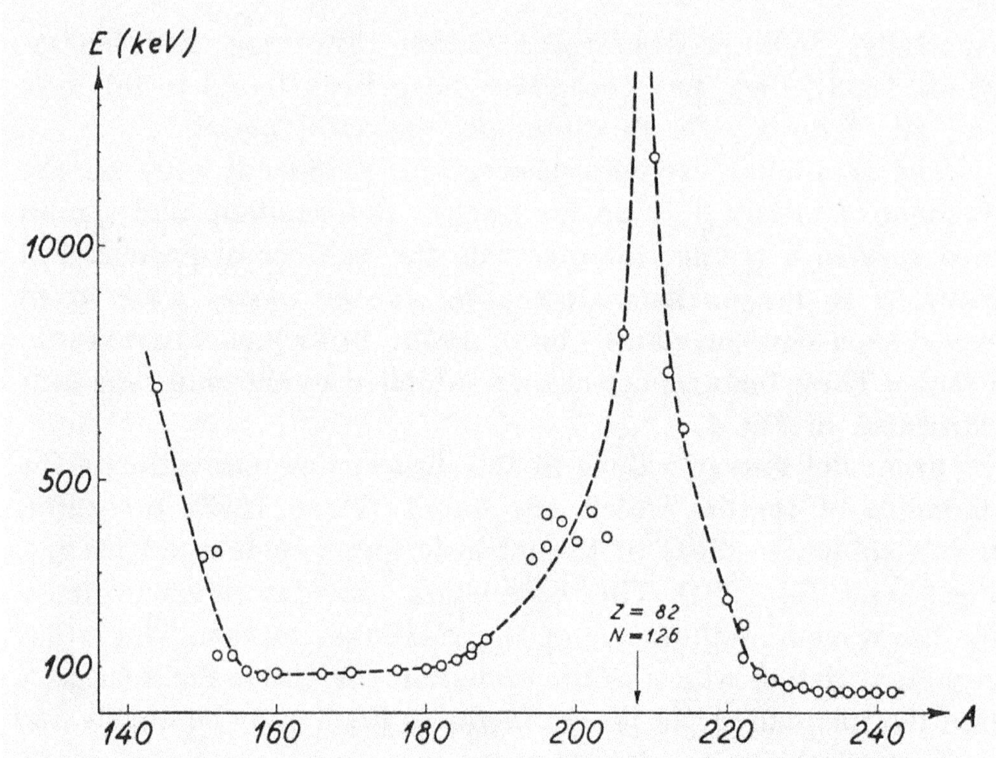

Fig. 4. *Energies of first excited states in even-even nuclei with $A > 140$.* The energy of the first excited state is plotted as a function of A. The evidence is consistent with a (2 +) assignment for all the listed levels. For the doubly closed-shell nucleus $_{82}Pb^{208}$ the first excited state at 2.61 MeV, which is not shown in the figure, has been recently found to be of (3 —) type (Elliot et al., 1954). The figure shows the expected decrease of the excitation energies as one moves away from closed-shell configurations, which is a consequence of the increasing nuclear deformation (cf. Fig. 1). In the regions $155 < A < 185$ and $A > 225$, where the energies are especially small, the levels are well represented as simple rotational states. With the approach to closed-shell configurations, the increasing frequency of rotation implies couplings to intrinsic nuclear types of motion which give the levels a more complex character. Finally, in the vicinity of closed-shell configurations, the nuclear excitation spectrum is entirely modified; in these regions, the first excited state may represent an excitation of individual particles, a collective vibrational mode, or a combination of both (cf. BM, § VI a, b, and c.i). The broken line in the figure indicates the smoothed-out variation of the excitation energies with A, but fluctuations from such an average behaviour are expected, especially in regions not too far removed from closed-shell configurations, where the character of the first excited state may vary from nucleus to nucleus. Moreover, the excitation energy is actually to be considered as a function of the two variables N and Z, rather than of A alone. It is only because of the simultaneous filling of neutron and proton shells at Pb^{208} that, for most of the nuclei considered, the deformation depends primarily on $N + Z = A$. The empirical data is taken from the compilations of Scharff-Goldhaber (1953), Stähelin and Preiswerk (1953), Hollander, Perlman, and Seaborg (1953). For the case of W^{184}, reference is made to Huus and Bjerregård (1953) and to McClelland, Mark, and Goodman (1954), and for Po^{210} to Hoff (1953).

especially, SCHARFF-GOLDHABER 1953; STÄHELIN and PREISWERK, 1953). Very few exceptions have been found to this rule and all of these refer to essentially spherical nuclei.

The rotational excitation energy (15) depends only on the moment of inertia, i. e. on the nuclear deformation, and should thus vary in a regular manner with the number of protons and neutrons in the nucleus, decreasing as one moves away from closed-shell configurations (FORD 1953; BOHR and MOTTELSON, 1953b). These features are clearly exhibited by the empirical data illustrated in Fig. 4.

From the energy values in this figure, one can estimate the moments of inertia, which are found (FORD 1953) to exhibit trends similar to those of the intrinsic quadrupole moments (cf. Fig. 1 and Eq. (28)). This correlation provides direct evidence for the wave-like character of the rotational motion. The rather regular variation with A of the excitation energies in Fig. 4 suggests that the fluctuations of the moments in Fig. 1 may be largely due to uncertainties in the Q-values.

The quantitative comparison between $\mathfrak{J}$ and Q_0 shows that the empirical Q_0-values are only about half as big as those obtained by means of (11) from the empirical $\mathfrak{J}$-values. This discrepancy may be associated with deviations from the simple spheroidal nuclear density, assumed in (11), and may also indicate a tendency for the protons to concentrate more strongly than the neutrons towards the center of the nucleus. Reasons for expecting such an effect have recently been discussed (JOHNSON and TELLER, 1954; L.WILETS, private communication).

In the region in Fig. 4, where the large nuclear deformations are expected, i. e. far removed from closed-shell configurations, the rotational energies are very small compared with the estimated vibrational energies, and in these regions one therefore expects the simple rotational spectrum to be rather accurately realized. In fact, for $155 < A < 185$ and for $A > 225$, the correction term (21) is estimated to be less than one per cent of the rotational energy. With the approach to the closed-shell configurations, the rotational frequencies increase and distortion terms are expected to become increasingly important. Finally, in the regions of the major shell closings, such as around Pb^{208}, the excitation spectrum acquires an essentially different character (cf. BM, § VIb and § VIc.i).

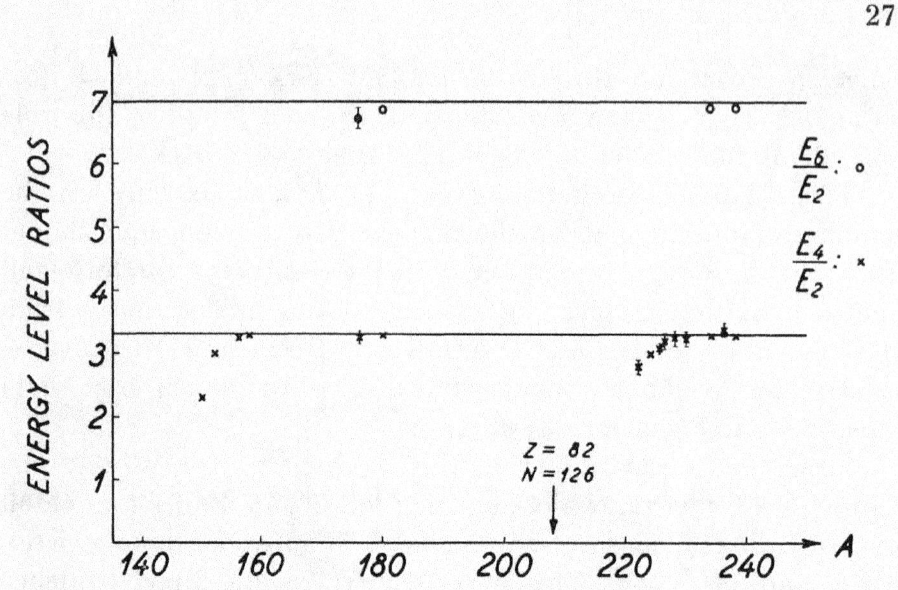

Fig. 5. *Energy ratios for rotational states in even-even nuclei.*
The ratio of the excitation energies of the second and third rotational states to that of the first is plotted against A. While the first rotational state is of (2 +) character, the second and third states are expected to be of (4 +) and (6 +) type, and the available empirical evidence is consistent with this assignment. The theoretical energy ratios given by (15) are shown by horizontal lines. These ratios are seen to be rather accurately realized for nuclei with large deformations, which are also characterized by small absolute values of the rotational energies (cf. Fig. 4). With the approach to closed-shell configurations, deviations from the simple rotational energy ratios are observed; these deviations have just the sign and order of magnitude expected for rotation-vibration interaction effects (cf. (21)). The estimated uncertainties in the measured energy ratios are indicated in the figure; where no error margin is shown, the uncertainty is believed to be less than one per cent. The empirical data is taken from
Arnold 1954 (Hf^{176}),
Asaro and Perlman, 1953 (Ra^{222}, $Th^{226,\,228,\,230}$, U^{236}),
Church and Goldhaber, 1954 ($Gd^{156,\,158}$),
Hibdon and Muehlhause, 1952 (Sm^{150}),
Keller and Cork, 1951 (Sm^{152}),
Mihelich, Scharff-Goldhaber, and McKeown, 1954 (Hf^{180}),
Newton and Rose, 1954 (Ra^{224}),
I. O. Newton and B. Rose, priv. com. (U^{234}, Pu^{238}),
Rasetti and Booth, 1953 (Ra^{226}).
A figure similar to the present was first given by Asaro and Perlman (1953).

These expectations are confirmed by the evidence on the position of the higher states in the spectrum. The second and third rotational states are expected to be of (4 +) and (6 +) character and, according to (15), to have the energies

$$E_4 = \frac{10}{3} E_2 \quad \text{and} \quad E_6 = 7\, E_2, \tag{22}$$

respectively. States of these spin and parity values have been observed to occur systematically for deformed nuclei, and the

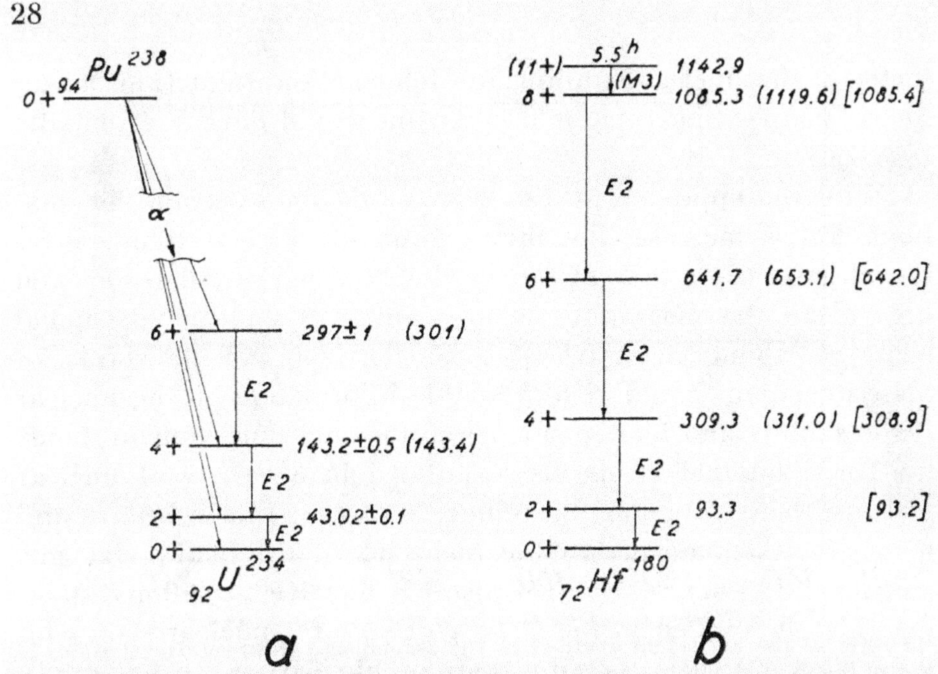

Fig. 6. *Examples of decay schemes involving rotational excitations of high order.* The rotational states occur systematically in the α-spectra of the heavy even-even nuclei (Asaro and Perlman, 1952, 1953; Rosenblum and Valadares, 1952). The most accurately measured rotational excitation energies in the α-spectra are those resulting from the decay of Pu^{238} (Fig. 6a) (J. O. Newton and B. Rose, to whom I am indebted for the permission to quote their results prior to publication). The measured energies are listed in keV and the numbers in parentheses are the theoretical values obtained from (15) by adjusting $\mathfrak{J}$ to give the observed value for E_2. The small deviations have the sign and order of magnitude expected of a rotation-vibration interaction effect (21), for a vibrational energy of about one MeV. The population of the states decreases rapidly with increasing energy and the intensities of the decay to states higher than (6 +) are too weak to be detected.

Fig. 6b shows the probable decay scheme for the 5.5 hour isomeric state in Hf^{180} formed by neutron capture in Hf^{179} (Goldhaber and Hill, 1952; Bohr and Mottelson, 1953b; Mihelich, Scharff-Goldhaber, and McKeown, 1954). The observed excitation energies are listed in keV and the numbers in parentheses are those obtained from (15) by determining $\mathfrak{J}$ from the observed value of E_2. The minor deviations from (15) are of the form (21), as seen from the numbers in square brackets, which represent a spectrum of the type $E_I = AI(I+1) - BI^2(I+1)^2$ with $A = 15.582$ keV and $B = 0.00703$ keV. These values of A and B, which give the best fit to the observed spectrum, correspond to $\hbar^2/\mathfrak{J} = 31.16$ keV and to a mean vibrational energy of about 2 or 3 MeV.

available data on the energy ratios $E_4:E_2$ and $E_6:E_2$ is shown in Fig. 5. It is seen that, in the regions where the well developed rotational spectrum is expected, the energy ratios are indeed very close to the values given by (22). The most precise energy determinations agree with the rotational ratios to within about one per cent. This agreement also confirms the expected axial symmetry of the nuclear equilibrium deformation. In fact, for asymmetric shapes, the rotational spectrum would deviate essentially from (15).

With the approach to $A = 208$, the deviations from the rotational ratios increase, but they exhibit the regular trend corresponding to (21); they also have the expected negative sign and are of just the magnitude corresponding to a mean vibrational energy of about one or two MeV. Further experimental studies of these deviations could yield valuable information on the nuclear deformability and its dependence on the nucleonic configurations.

The rotational levels are populated in a variety of nuclear processes; in particular, the states up to (6 +) have been found to occur systematically in the α-spectra of the heavy elements (Asaro and Perlman, 1953; cf. also Fig. 6a). High rotational states may also be excited in the β- or γ-decay of a nuclear state with a high spin value. As an example, Fig. 6b shows the decay scheme for an isomeric state in Hf^{180}, which appears to be of (11 +) type[1] and to decay to the (8 +) rotational excitation of the ground state. This state then decays by the emission of a cascade of four γ-rays with energy ratios closely equal to the values 15:11:7:3 given by (15). The small deviations from these ratios have the sign and order of magnitude expected of a rotation-vibration interaction, and have just the I-dependence given by (21). Thus, with an accuracy of about one part in a thousand, the observed levels can be fitted to a rotational spectrum, when due allowance is made for the finite but small rotational frequency.

The rotational states also show up in a striking way in the Coulomb excitation process (cf. § Vc). By this method, rotational states in the odd-A nucleus $_{73}Ta^{181}$ have recently been identified (Huus and Zupančič, 1953; Huus and Bjerregård, 1953; cf. Fig. 7). The ground state of Ta^{181} is known to have $I_0 = 7/2$, and two strongly excited states have been observed with the energies 137 keV and 303 keV, respectively. They have been interpreted as the 9/2 and 11/2 two first rotational states (cf. (19)),

[1] Some uncertainty remains in the multipole order of the 57 keV isomeric transition, which may be smaller than 3. If this is the case, the spin assignment of the isomeric state is to be correspondingly decreased.

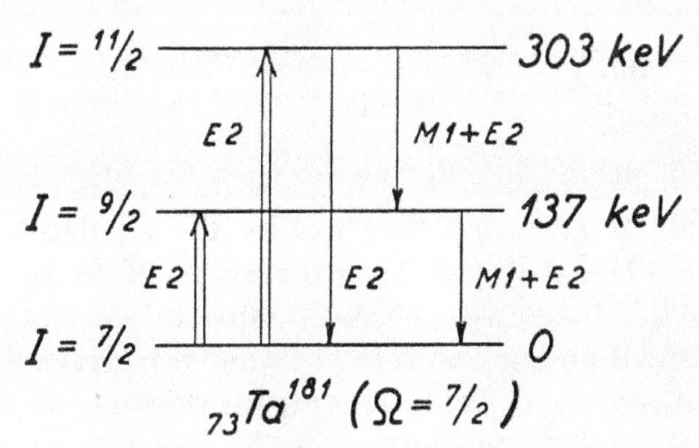

Fig. 7. *Rotational spectrum of* Ta^{181}.

The best studied rotational spectrum of an odd-A nucleus is that of ${}_{73}Ta^{181}$. By means of the Coulomb excitation process (double arrows in the figure), the two first rotational states have been excited. The ground state spin of Ta^{181} is known from spectroscopic data to be $I = 7/2$ (cf., e. g., Mack 1950), and the first and second rotational states are thus expected to have $I = 9/2$ and $11/2$, respectively. These spin assignments have recently been confirmed by angular correlation measurements (McGowan 1954; Eisinger et al., 1954).

The energies and radiative decays of the rotational states are shown in the figure. The study of these processes (Huus and Zupančič, 1953; Huus and Bjerregård, 1953) have provided a number of tests of the theory of rotational states and yielded various information on the intrinsic nuclear properties. We list the quantities on which empirical evidence is available.

1) $E_{11/2}$: $E_{9/2} = 2.21 \pm 0.02$ (theor. value 2.22).
2) $E_{9/2} = 137$ keV (theor. estimate, using $\mathfrak{J}$ of Hf^{180}, gives $E_{9/2} \approx 140$ keV).
3) $B(E2;\ 7/2 \rightarrow 9/2) : B(E2;\ 7/2 \rightarrow 11/2) = 3.8 \pm 0.7$ (theor. value 3.89).
4) $B(E2;\ 7/2 \rightarrow 9/2)$ yields $|\,Q_0\,| \approx 7 \times 10^{-24}$ cm^2, in accordance with the deformation expected from the data in Table I; spectroscopic determinations (Schmidt 1943; Brown and Tomboulian, 1952) have led to an average Q_0-value of about 14×10^{-24} cm^2, but with a rather large uncertainty.
5) $B(M1;\ 11/2 \rightarrow 9/2) : B(M1;\ 9/2 \rightarrow 7/2) = 2.5 \pm 1.0$ (theor. value 1.53).
6) $B(M1;\ 11/2 \rightarrow 9/2)$ together with $\mu = 2.1$ (Brown and Tomboulian, 1952) yield $g_\Omega \approx 0.75$ and $g_R \approx 0.25$.

and later measurements (McGowan 1954; Eisinger et al., 1954) have confirmed these spin assignments. The observed energy ratio 2.21 ± 0.02 agrees closely with the value $20/9$ obtained from (18). The absolute values of the energies depend on the moment of inertia, but may be estimated from the rotational energies of neighbouring even-even nuclei, if one neglects the change in deformation caused by the last odd particle. Thus, the spectrum of ${}_{72}Hf^{180}$ (cf. Fig. 6b) gives a value for $\mathfrak{J}$ which leads to an energy of 140 keV for the first rotational excitation in ${}_{73}Ta^{181}$, in good agreement with the observed value.

Rotational states also appear to have been observed in a number of other odd-A nuclei (cf., e. g., Rasmussen 1954; Temmer and Heydenburg, 1954; McClelland, Mark, and Goodman,

1954)[1]. No evidence is at present available on rotational states in odd-odd nuclei.

The present discussion has been limited to nuclei with $A > 140$, which provide the main evidence on nuclear rotational spectra. Also for lighter nuclei, however, rotational states are expected in regions removed from closed-shell configurations. It is known that in these regions the energies of the first excited (2+) states in even-even nuclei have pronounced minima (cf. SCHARFF-GOLDHABER 1953; STÄHELIN and PREISWERK, 1953) and are considerably smaller than the estimated vibrational energies. There is also some direct evidence for the existence in these regions of sequences of states having the spins and, approximately, the energy ratios characteristic of rotational spectra (cf., e.g., Mg^{24} (HEDGRAN and LIND, 1952); Xe^{130} (CAIRD and MITCHELL, 1954)).[2]

It is also added that we have here considered only the rotational spectrum based on the nuclear ground state. For each intrinsic nuclear excitation, which may correspond to a change of binding state of individual particles, or to a collective vibration, one expects a new rotational band of similar properties as that associated with the nuclear ground state.

With increasing excitation energy, however, the perturbations in the rotational motion become of growing importance due to increasing nuclear level density. Thus, in the region of excitation involved in nuclear reaction processes, the stationary states acquire an essentially greater complexity, corresponding to the characteristic properties of the compound nucleus formed in such processes. (Cf. BOHR and KALCKAR, 1937; HILL and WHEELER, 1953; BM, § VI d and Appendix V).

[1] Note added in proof: Recent Coulomb excitation studies by G. TEMMER and N. HEYDENBURG, and by T. HUUS (to whom I am indebted for private communications of results of their work) have provided a rather extensive body of data, confirming the systematic occurrence of rotational spectra in odd-A nuclei in regions of large deformations. As in Ta^{811}, two states are strongly excited in the process, and the ratios of the excitation energies are found to agree closely with those obtained from (18).

[2] An analysis of this data is being prepared by B. R. MOTTELSON and the writer.

V. Electromagnetic Moments and Transitions.

The collective rotation of the large nuclear quadrupole gives rise to electric quadrupole transitions between rotational levels, of an intensity several orders of magnitude larger than for transitions between single-particle states. Thus, the measurement of transition probabilities is an important tool in the identification of rotational excitations.

Due to the simplicity of the rotational motion, the observed transition probabilities of electric as well as magnetic type can be directly related to the static electric and magnetic moments, and together with these provide information on the intrinsic electromagnetic properties of the deformed nucleus.

Expressions are given below, first for the static moments and, subsequently, for the transition probabilities; the latter can be determined either from a study of radiative transitions or from a measurement of cross sections for Coulomb excitation. Finally, the information which can be obtained from the available empirical evidence is summarized.

a) Static moments.

The magnetic dipole moment of the rotating nucleus is composed of two parts, the "intrinsic" moment, associated with the angular momentum Ω of the particle structure, and the collective moment, associated with the rotational angular momentum $\vec{R}$. For the coupling scheme illustrated in Fig. 3, one finds for the total nuclear magnetic moment

$$\mu = g_{\Omega} \frac{\Omega^2}{I+1} + g_R \frac{I(I+1)-\Omega^2}{I+1}, \qquad (23)$$

where g_Ω and g_R denote the gyromagnetic ratios for the intrinsic and collective angular momenta, respectively. The moment μ, as well as the g-factors, are given in units of the nuclear magneton

$$\mu_N = \frac{e\hbar}{2\,mc}. \tag{24}$$

The intrinsic g-factor is a characteristic of the binding state of the last odd particle, and depends especially on the relative contributions to Ω of the orbital and spin angular momenta of the particle. The collective g-factor depends on the electric current associated with the collective motion. For a proton to neutron density ratio constant throughout the nucleus, a value of about

$$g_R \approx \frac{Z}{A} \tag{24a}$$

is expected.

For an even-even nucleus with $\Omega = 0$, one obtains from (23) simply

$$\mu = g_R I. \tag{25}$$

For the ground state of an odd-A nucleus, we have $I_0 = \Omega$, except for $\Omega = 1/2$, and (23) gives

$$\mu = g_\Omega \frac{I_0^2}{I_0 + 1} + g_R \frac{I_0}{I_0 + 1}. \tag{26}$$

In the special case of $\Omega = 1/2$, for which an anomalous rotational spectrum occurs (cf. (20)), also the magnetic moment depends more specifically on the particle state, and expressions (23) and (26) need modifications (cf. BM, § IVb and Addendum to Chapters IV and V).

The nuclear quadrupole moment Q, defined by (1), is related to the intrinsic quadrupole moment Q_0 (cf. (2)) which characterizes the actual nuclear deformation. For a rotational state with quantum numbers Ω and I, one finds (cf. BM, § Vb)

$$Q = \frac{3\,\Omega^2 - I\,(I+1)}{(I+1)\,(2\,I+3)}\,Q_0 \tag{27}$$

which, for the nuclear ground state ($I_0 = \Omega$), becomes

$$Q = \frac{I_0\,(2\,I_0 - 1)}{(I_0 + 1)\,(2\,I_0 + 3)}\,Q_0. \tag{28}$$

The magnitude of Q is always smaller than that of Q_0, and Q vanishes for $I = 0$ or $1/2$. Indeed, any quantum state with total angular momentum 0 or 1/2 possesses a spherically symmetric charge distribution, even though the system may have an intrinsic deformation.

Expression (27) still holds for the case of $\Omega = 1/2$, but the ground state may not then have $I_0 = \Omega$.

It is noted that the expressions for nuclear moments given in this section, as well as the transition probabilities given in the following one, are valid only for the strongly deformed nuclei which possess a well developed rotational spectrum. For less deformed nuclei, the coupling between the rotational and intrinsic degrees of freedom, which modifies the rotational spectrum (cf. (21)), also implies corrections to the moments and transition probabilities.

b) Radiative transition probabilities.

Successive levels in the rotational spectrum have a spin difference of one or two units and the same parity, and the radiative transitions are therefore mainly of $M1$ (magnetic dipole) or $E2$ (electric quadrupole) type.

The probability per second for a γ-transition between two states with the energy difference ΔE may be written (cf., e. g., Blatt and Weisskopf, 1952; we follow the notation used in BM, Chapter VII)

$$T(M1) = \frac{16\pi}{9}\frac{1}{\hbar}\left(\frac{\Delta E}{\hbar c}\right)^3 B(M1) \qquad (29)$$

and

$$T(E2) = \frac{4\pi}{75}\frac{1}{\hbar}\left(\frac{\Delta E}{\hbar c}\right)^5 B(E2) \qquad (30)$$

for the two types of transitions in question. The quantities $B(M1)$ and $B(E2)$ are the reduced transition probabilities which do not involve ΔE, but only the square of the nuclear matrix elements of the magnetic dipole and electric quadrupole operators, respectively.

In an even-even nucleus, where the sequence of states in the rotational spectrum is given by (17), an excited level decays by a cascade of pure $E2$ radiation. The transition probabilities may be expressed in terms of the intrinsic nuclear quadrupole moment Q_0, defined by (2) and, for the transition $I+2 \rightarrow I$, one finds (BM, VII. 17)

$$B(E2) = \frac{15}{32\pi} e^2 Q_0^2 \frac{(I+1)(I+2)}{(2I+3)(2I+5)}. \tag{31}$$

For values of Q_0 of the order of those indicated by the spectroscopically measured large nuclear quadrupole moments, the value of $B(E2)$ exceeds that of a single-particle transition by a factor of about a hundred.

In an odd-A nucleus, or an odd-odd nucleus, the spin difference between successive states (cf. (19)) is normally one unit (except in the anomalous case of $\Omega = 1/2$ (cf. (20)), and an excited level decays by emission of mixed $M1$ and $E2$ radiation.

The magnetic transition probabilities in the rotational spectrum can be expressed in terms of the intrinsic and collective g-factors, which also determine the static magnetic moment (cf. (23)). For the decay $I+1 \rightarrow I$, one finds (BM, VII. 20), for $\Omega \neq 1/2$,

$$B(M1) = \frac{3}{4\pi}\left(\frac{e\hbar}{2mc}\right)^2 (g_\Omega - g_R)^2 \frac{\Omega^2 (I+1-\Omega)(I+1+\Omega)}{(I+1)(2I+3)}. \tag{32}$$

This expression vanishes for $g_\Omega = g_R$; the total magnetic moment has then the direction of $\vec{I}$ and is, therefore, a constant of the motion, which cannot give rise to emission of radiation.

The intensity of the $E2$ radiation in the rotational transition $I+1 \rightarrow I$ is given by (BM, VII. 18)

$$B(\mathrm{E}2) = \frac{15}{16\pi} e^2 Q_0^2 \frac{\Omega^2 (I+1-\Omega)(I+1+\Omega)}{I(I+1)(2I+3)(I+2)}. \tag{33}$$

In a single-particle transition of $\Delta I = 1$ and no parity change, and with an energy ΔE of the order of those here in question, the intensity of $M1$ radiation is expected to exceed that of $E2$ by a factor of the order of 10^3 or 10^4. However, the large collective nuclear quadrupole moments imply that $E2$ radiation may effectively compete with $M1$ in the rotational transitions.

One thus also expects an appreciable intensity of cross-over transitions of the type $I+2 \rightarrow I$. These are of pure $E2$ character with (BM, VI. 19)

36

$$B(E2) = \frac{15}{32\pi} e^2 Q_0^2 \frac{(I+1-\Omega)(I+1+\Omega)(I+2-\Omega)(I+2+\Omega)}{(I+1)(2I+3)(I+2)(2I+5)}, \quad (34)$$

which reduces to (31) for $\Omega = 0$.

In the special case of $\Omega = 1/2$, where an anomalous sequence of states may occur, the $E2$ transition probabilities are still given by (33) and (34), but the $M1$ transition probabilities depend on more detailed features of the particle state.

c) Coulomb excitation cross sections.

An alternative way of determining electric transition probabilities in the nucleus has recently become available in the Coulomb excitation process, in which the nucleus is excited by the electric field of impinging particles (McClelland and Goodman, 1953; Huus and Zupančič, 1953). When the projectile has insufficient energy to penetrate the Coulomb barrier, the cross section for excitation depends only on the electrostatic interaction and can be expressed in terms of the same nuclear matrix elements which determine the radiative transition probabilities of electric multipole type. The large quadrupole matrix elements in the rotational spectrum make the Coulomb excitation process especially suited for the study of rotational excitations.

For the velocities in question, the projectile may be considered, to a first approximation, as moving along a classical trajectory, and the excitation probability may be determined from the time-dependent electric field with which the particle acts on the nucleus (Ter-Martirosyan 1952). For a quadrupole excitation, one obtains in this way (cf. also BM, Ap. VI)

$$\sigma_{\text{exc}} = \frac{2\pi^2}{25} \frac{1}{Z_2^2 e^2} \left(\frac{Mv}{\hbar}\right)^2 B(E2)\, g_2(\xi), \quad (35)$$

where $Z_2 e$ is the charge of the target nucleus, v the relative velocity, and M the reduced mass of the collision.

The quantity $B(E2)$ is the reduced $E2$ transition probability. For the excitation from a state i to a state f, the value of B is related to that of the inverse process, the radiative decay from f to i, by the equation

$$B(i \to f) = B(f \to i) \frac{2I_f + 1}{2I_i + 1}, \quad (36)$$

where I_i and I_f are the spins of the two combining states. For rotational excitations, the value of $B(E2)$ to be inserted in (35) may thus be obtained from (31), (33), and (34), taking into account (36).

Finally, the function g_2 depends on the dimensionless parameter

$$\xi = \frac{Z_1 Z_2 e^2}{\hbar}\left(\frac{1}{v_f} - \frac{1}{v_i}\right) \approx \Delta E \frac{Z_1 Z_2 e^2}{\hbar M v^3}, \tag{37}$$

where $Z_1 e$ is the charge of the projectile and where v_i and v_f are the relative velocities in the initial and final stages of the collision. The velocity v is the mean of v_i and v_f, and it is assumed that the excitation energy ΔE is small compared with the kinetic energy of the incident particle. The quantity ξ represents the ratio between the collision time and the nuclear period. The function g_2 can be expressed in terms of integrals over the trajectories of the impinging particle, and has been evaluated numerically (ALDER and WINTHER, 1953). For $\xi \ll 1$, the value of g_2 is of order unity, but for $\xi \gtrsim 1$, the collisions become approximately adiabatic and g_2 decreases rapidly with increasing ξ.[1]

d) Empirical data.

For the even-even nuclei, a considerable number of transition probabilities have been determined for transitions between the ground state and the first excited (2 +) state. The data are listed in Table I and are obtained partly from lifetime determinations and partly from Coulomb excitation cross sections.

The intensity of the transitions is indicated by the quantity F in column three, which represents the ratio of the observed transition probability to that expected for a particle transition with corresponding spin and parity change. The largeness of the F-factors clearly exhibits the collective nature of the excitations (GOLDHABER and SUNYAR, 1951; BOHR and MOTTELSON, 1952; 1953a) and is the more striking, since rotational transitions are the only ones hitherto found in nuclei with transition probabilities appreciably exceeding those of single-particle transitions.

The transitions in Table I are of pure $E2$ type, and from the

[1] A review article containing a discussion of the theory of Coulomb excitation, as well as of experimental techniques and results, and of their interpretation, is being prepared by K. ALDER, A. BOHR, T. HUUS, B. R. MOTTELSON, and A. WINTHER.

TABLE I. Rotational Transitions in Even-Even Nuclei.

Nucleus	E(keV)	F	Q_0 (10^{-24} cm^2)		
			rad. decay	Coul. exc.	spectroscopic
$_{66}Dy^{160}$	85	80	7	..	..
$_{68}Er^{166}$	80	100	8	..	~20 ($_{68}Er^{167}$)
$_{70}Yb^{170}$	84	100	8	..	11 ($_{70}Yb^{173}$)
$_{72}Hf^{176}$	89	90	8	..	14 ($_{71}Lu^{175}$)
$_{74}W^{182}$	100	80	7	7	..
$_{74}W^{184}$	113	80	..	7	14 ($_{73}Ta^{181}$)
$_{74}W^{186}$	124	80	..	7	8 ($_{75}Re^{185}$)
$_{76}Os^{186}$	137	50	6	..	7 ($_{75}Re^{187}$)

The table lists the rotational states in even-even nuclei whose lifetimes or Coulomb excitation cross sections have been measured. The quantity F in the third column gives the ratio of the observed transition probability to that calculated for a single-particle transition (cf., e. g., BM, § VII b.i). From the observed transition probability the intrinsic quadrupole moments listed in columns four and five have been obtained by means of the expressions given in § V b and c. The last column gives the Q_0-values obtained by means of (28) from spectroscopic Q-determinations in neighbouring odd-A nuclei.

The energies in column two are taken from SCHARFF-GOLDHABER (1953), HUUS and BJERREGÅRD (1953), and McCLELLAND, MARK, and GOODMAN (1954). For the radiative transitions, the empirical lifetimes have been taken from GOLDHABER and HILL (1952). In order to determine the transition probability for emission of γ-radiation, correction must be made for the internal conversion effect. For the K-conversion coefficient, α_K-values obtained by extrapolation of those given by ROSE et al. (1951) were used, while the L-conversion coefficients α_L have been estimated from the empirical relationship for $\alpha_K:\alpha_L$ (GOLDHABER and SUNYAR, 1951). A re-evaluation of this data (B. MOTTELSON, private communication) has led to somewhat larger total conversion coefficients than those previously assumed, and the F- and Q_0-values in Table I are therefore somewhat smaller than those listed in Table XXVII of BM. The Coulomb excitation cross sections are obtained from HUUS and ZUPANČIČ (1953). Tables of spectroscopic Q-values may be found in BM (Addendum to Chapters IV and V).

lifetime or excitation cross section, the intrinsic nuclear quadrupole moment Q_0 can be obtained by means of (30), (31), or (31), (35), (36), respectively. The values of Q_0 are listed in columns four and five, and the results obtained by the two methods are seen to be in good agreement. Moreover, the Q_0-values show the expected trend, decreasing regularly with the approach to closed-shell configurations. This trend of the nuclear deformations corresponds to the increasing rotational energies listed in the second column.

The estimated Q_0-values may be compared with the static quadrupole moments obtained from spectroscopic measurements. Even-even nuclei have spin zero in their ground state and, thus, do not give rise to a hyperfine structure of the atomic spectrum[1], but one may attempt to use data from neighbouring odd-A nuclei. The Q_0-values in the last column of Table I are obtained in this way, by means of the relation (28).

It is seen that the static $E2$ (quadrupole) moments yield deformations of the same order of magnitude as those obtained from $E2$ transition probabilities. The former are somewhat larger than the latter, but the difference may not be significant in view of the experimental uncertainty in the spectroscopic Q-values, which may also be responsible for the less regular trend of the Q_0-values in the last column (cf. above, p. 26). Part of the difference may also be due to a tendency for odd-A nuclei to have larger deformations than neighbouring even-even nuclei.

Information on the rotational transition probabilities in odd-A nuclei has been obtained from the Coulomb excitation study of $_{73}Ta^{181}$ (Huus and Zupančič, 1953; Huus and Bjerregård, 1953). The ground state of Ta^{181} is known to have $I_0 = 7/2$, and an $E2$ Coulomb excitation process can thus excite the two first rotational states with $I = 9/2$ and $11/2$ (cf. Fig. 7). Corresponding to the transitions in neighbouring even-even nuclei (cf. Table I), the cross sections are found to be about a hundred times larger than for a particle excitation. For the $7/2 \rightarrow 9/2$ excitation, one derives, by means of (33), (35), and (36), the deformation $|Q_0| \approx 7 \times 10^{-24}\,\text{cm}^2$, which is in good agreement with the trend of deformations in Table I. Spectroscopic determinations of Q (Schmidt 1943; Brown and Tomboulian, 1952) have yielded a Q_0-value for Ta^{181}, which is positive, and whose magnitude is on the average about twice that obtained from the excitation cross section. However, the Q-values derived from the hyperfine structure of different lines in the Ta-spectrum show very large variations.

The ratio of the $7/2 \rightarrow 11/2$ excitation cross section to the $7/2 \rightarrow 9/2$ cross section is independent of Q_0, and from (33), (34), and (36) one obtains a ratio of 9:35 for the reduced transition probabilities. This is in accordance with the experimentally determined ratio of about 1:4.

[1] The possibility of determining the intrinsic quadrupole moments of even-even nuclei from the spectrum of μ-mesonic atoms has recently been suggested by Wilets (1954).

The radiation emitted by the Coulomb excited states in Ta^{181} has also been studied in some detail (HUUS and ZUPANČIČ, 1953; HUUS and BJERREGÅRD, 1953). The 9/2 state decays by a mixed $M1+E2$ transition, and the 11/2 state may decay either directly to the ground state by a pure $E2$ transition or to the 9/2 state by a mixed $M1+E2$ transition (cf. Fig. 7). The $E2$ intensities of the various transitions are given by (33) and (34), using the Q_0-value determined from the absolute value of the excitation cross sections. The $M1$ intensities can thus be obtained from relative measurements, and have been determined, for the $11/2 \rightarrow 9/2$ transition, from the branching ratio between this transition and the $11/2 \rightarrow 7/2$ cross-over. For the $9/2 \rightarrow 7/2$ transition, the $M1$ intensity has been estimated from the known $E2$ intensity and the internal conversion coefficients which yield the $M1$: $E2$ ratio. This ratio has also been determined by angular correlation methods (McGOWAN 1954). The reduced $M1$ transition probablities of the two decays obtained in this way are found to have a ratio of 2.5 ± 1.0, consistent with the theoretical value 1.53 given by (32).

Moreover, the absolute $M1$ intensities determine the quantity $(g_\Omega - g_R)$ which is found from (29) and (32) to be ≈ 0.5.[1] From the empirical value $\mu = 2.1$ (BROWN and TOMBOULIAN, 1952) for the static magnetic moment of the Ta^{181} ground state, one can then determine the g-factors separately, by means of (26), and one finds $g_\Omega \approx 0.75$ and $g_R \approx 0.25$. The latter value is a significant characteristic of the collective nuclear motion. It is somewhat smaller than the estimate $\frac{Z}{A} \approx 0.40$ (cf. (24a)) for the uniformly charged nucleus; this may suggest a relative concentration of protons towards the center of the nucleus, as may also be indicated by the comparison of $E2$ moments with the moments of inertia (cf. above, p. 26). Additional measurements of $M1$ transition probabilities in rotational spectra, together with magnetic moment determinations, could further elucidate this point.

[1] The $M1$ intensity does not give the sign of $(g_\Omega - g_R)$; however, the assumption of $g_\Omega - g_R \approx -0.5$ leads to the less probable g_R-value of 1.0.

VI. Concluding Remarks.

In the foregoing, evidence has been presented for the systematic occurrence, in a large class of nuclei, of rotational spectra characterized by numerous regularities in spins, excitation energies, and transition probabilities. The existence of these states clearly illustrates how the nuclear dynamics is governed by the interplay between the individual-particle and collective types of motion. The nucleus acts like a deformed body whose rotation is generated by a wave-like collective motion of the particles. At the same time, the occurrence of the deformations is directly related to the nuclear shell structure and shows that the binding of the individual particles is dependent upon the shape of the whole nucleus.

The significance of the existence of nuclear rotational spectra goes beyond the implications for the general structure of the nucleus. The great simplicity and regularity of the rotational states place them among the best understood nuclear levels; indeed, for hardly any other states have quantitative predictions of excitation energies or transition probabilities been possible. Thus, empirical data regarding their occurrence in α- and β-decay, γ-processes, nuclear reactions, etc. is relatively easy to interpret and may aid in the exploration of the mechanism involved in these various nuclear phenomena.

Appendix:

On the Collective Flow in a Quantum Mechanical Many-Particle System.[1]

The transformation of the equations of motion of a many-particle system to the form applied in the unified collective and individual-particle treatment of the nucleus may be considered as a co-ordinate transformation which introduces collective co-ordinates, in addition to the intrinsic co-ordinates describing the nuclear shell structure (cf. BM, § IIa). This procedure is briefly indicated in the following, for the degrees of freedom associated with the rotational spectrum.

The rotational motion can be described in terms of a collective flow, defined by means of the collective co-ordinates. This flow obeys equations of the hydrodynamical type, from which, in particular, the effective moment of inertia for the rotational spectrum can be determined.

a) Collective and intrinsic co-ordinates.

We consider a system of particles in which the binding forces give rise to a non-spherical equilibrium shape, and we assume the density fluctuations about equilibrium to be small compared to the mean deformation.

For such a deformed system, an approximate separation between intrinsic types of motion and a collective rotational motion is expected to be possible (cf. p. 24 above). In order to give the equations of motion a form which allows such a separation, one must introduce an appropriate set of co-ordinates, consisting

[1] The writer is indebted to Drs. K. Alder, A. R. Edmonds, B. R. Mottelson, and M. Trocheris for illuminating discussions on the topics of this appendix.

partly of intrinsic co-ordinates, and partly of collective co-ordinates specifying the orientation of the system as a whole.

The particular choice of these co-ordinates depends on the interactions between the particles, and we shall illustrate the situation by considering two extreme types of systems, a quantum shell structure and a rigid body.

In a shell structure like the nucleus, where the particles move freely among each other, the angular co-ordinates characterizing the orientation of the system must be symmetric functions of the particle co-ordinates. For example, one may use the principal axes of the ellipsoid of inertia to specify an "intrinsic" or "body-fixed" co-ordinate frame, and introduce the three Eulerian angles Θ_i $(i = 1, 2, 3)$ to describe the orientation of this intrinsic frame with respect to a fixed co-ordinate frame.

To obtain these Θ_i in terms of the particle co-ordinates, we denote by $a_{\mu\nu}(\Theta_i)$ the transformation matrix from the Cartesian position co-ordinates $x_{p\nu}$ of a particle p in the fixed system to the co-ordinates $x'_{p\mu}$ in the intrinsic system, so that

$$x'_{p\mu} = \sum_{\nu} a_{\mu\nu}(\Theta_i)\, x_{p\nu}. \tag{I}$$

Then, the angles Θ_i are given by the implicit equations[1]

$$\sum_{p,\nu,\nu'} a_{\mu\nu}(\Theta_i)\, a_{\mu'\nu'}(\Theta_i)\, x_{p\nu} x_{p\nu'} = \delta_{\mu\mu'} \sum_{p} (x'_{p\mu})^2. \tag{II}$$

This particular definition of Θ_i is associated with the quadrupole mass tensor, and thus applies only to a system whose deformation is of pure quadrupole type. Moreover, even for a pure quadrupole deformation, the equation (II) is not the most general, since it assumes a definite weighting of the particles, according to their distance from the origin. The proper definition of the Θ_i for a given system depends on the density distribution, and we shall discuss later the conditions under which (II) applies.

The collective orientation angles must be chosen quite differently for a rigid (or quasi-rigid) system, where the particles are located near certain equilibrium positions. Thus, in a molecule, the separation of the rotation from the vibratory motions is

[1] This definition of the Θ_i is equivalent to the definition (BM, § II.2) of the deformation parameters $\alpha_{2\mu}$, from which, in turn, the orientation angles are obtained by equation (A, 10).

achieved by introducing the orientation angles through the equations (cf., e.g., NIELSEN 1951)

$$\sum_p M_p(\vec{\xi}_p \times \vec{x}_p) = 0, \tag{III}$$

where M_p is the mass of the pth nucleus, and $\vec{\xi}_p$ its equilibrium position. The components of $\vec{\xi}_p$ along the intrinsic co-ordinate axes are constants, and the equation (III) expresses the condition that the angular momentum in the intrinsic co-ordinate system vanishes in first order. Since $\vec{\xi}_p$ differs from particle to particle, the Θ_i given by (III) are essentially unsymmetric in the particle co-ordinates.

The intrinsic co-ordinates we denote by q_α. In a rigid body, they are to be chosen as the normal co-ordinates characterizing the proper modes of vibration of the system. In a quantum shell structure, it is convenient to use the individual-particle co-ordinates $x'_{p\mu}$ in the body-fixed system, together with suitable spin co-ordinates. Actually, the $x'_{p\mu}$ are not fully independent variables in view of (II), but the bonds between them can often be neglected, corresponding to the fact that the system for many purposes behaves as though it had the degrees of freedom associated with all the co-ordinates $x'_{p\mu}$, as well as with the Θ_i.

In order to illustrate this point, we may recall that, even in the static spherical shell model, one has to do with collective degrees of freedom, associated with the center of gravity motion. The appropriate set of co-ordinates thus comprises the position of the center of gravity, in addition to the individual-particle co-ordinates relative to this center. Although the latter are not fully independent, it is still convenient to treat the system in terms of all the individual-particle co-ordinates plus the collective co-ordinates. In fact, the system possesses states associated with a translation of the center of gravity as well as states associated with each of the excitation modes of the shell structure. The bonds between the individual-particle co-ordinates imply that the shell model wave functions do not rigorously form an ortho-normal set, but the consequences are negligible, for many purposes, provided the number of particles is large. Only for states involving the simultaneous excitation of an appreciable fraction of all the particles would the bonds become essential.

Similarly, in the deformed rotating shell structure, the bonds between the individual-particle co-ordinates relative to the intrinsic co-ordinate frame become important only if the number of excited particles is sufficient to influence appreciably the shape and orientation of the nucleus.

b) The transformation of the Hamiltonian.

In order to transform the equations of motion to the co-ordinates q_α, Θ_i, we note that the velocity operator of a particle of mass m, when acting on a wave function of the type

$$\Psi = \Psi(q_\alpha, \Theta_i), \tag{IV}$$

takes the form

$$\vec{v}_p = \frac{\hbar}{im}\vec{\nabla}_p = \vec{v}_p^{(r)} + \vec{v}_p^{(c)}, \tag{V}$$

where

$$\vec{v}_p^{(r)} = \frac{\hbar}{im}\sum_\alpha (\vec{\nabla}_p q_\alpha)\frac{\partial}{\partial q_\alpha} \tag{VI}$$

is a relative velocity, and

$$\vec{v}_p^{(c)} = \frac{\hbar}{im}\sum_i (\vec{\nabla}_p \Theta_i)\frac{\partial}{\partial \Theta_i} \tag{VII}$$

a collective velocity.

The potential energy of the system is a function of the q_α only, while the kinetic energy may be obtained from (V). The nuclear Hamiltonian is then transformed into an expression of the type

$$H = H_{\text{intrins}}(q_\alpha) + H_{\text{rot}}(\Theta_i) + H_{\text{coupl}}(q_\alpha, \Theta_i). \tag{VIII}$$

The first term is the intrinsic energy which, for a shell structure, represents essentially the energy of the particles in a deformed field of fixed orientation. The second term in (VIII) is the collective rotational energy,

$$H_{\text{rot}} = \sum_p \frac{1}{2} m (\vec{v}_p^{(c)})^2, \tag{VIIIa}$$

while the last term in (VIII) represents the coupling between the intrinsic motion and the rotation, arising from the non-static character of the field in which the particles are bound. Thus, the

Hamiltonian has the general form characteristic of the unified nuclear model (A; BM; cf. the strong coupling representation).

The rotational energy (VIII a), as a function of the Θ_i and their associated momenta, has the same structure as for a rigid body, but the effective moments of inertia depend on the choice of the Θ_i as functions of the $\vec{x}_p$. Thus, the definition (II), appropriate to a shell structure, leads to

$$\mathfrak{J}_{z'} = \frac{m\left(\sum_p (x_p'^2 - y_p'^2)\right)^2}{\sum_p (x_p'^2 + y_p'^2)} \tag{IX}$$

for the moment of inertia with respect to the intrinsic z'-axis; x'_p and y'_p are the co-ordinates of the pth particle with respect to the intrinsic x'- and y'-axes. The expression (IX) is approximately a constant, provided the fluctuations in the density distribution are small compared with the deformation.

The moment of inertia given by (IX) is proportional to the square of the deformation, as found in § III above for the wave-like rotations. The dependence of $\mathfrak{J}$ on the deformation is characteristic of the symmetric co-ordinates $\Theta_i(\vec{x}_p)$.

If the deformation approaches cylindrical symmetry, the moment of inertia about the symmetry axis becomes very small and the energies associated with a rotation about this axis very large. At the same time, the angular momentum about this axis becomes a constant of the motion, denoted in § IV above by Ω[1]. Thus, the low-lying rotational states of an axially symmetric nucleus are characterized by fixed Ω and are associated only with rotations about axes perpendicular to the symmetry axes.

In the case of a rigid body, for which Θ_i is given by (III), the moments of inertia occurring in H_{rot} are given by the familiar expression

$$\mathfrak{J}_{z'} = \sum_p M_p (\xi_p^2 + \eta_p^2) \tag{X}$$

and its cyclic permutations. The components of $\vec{\xi}_p$ along the intrinsic x'- and y'-axes are denoted by ξ_p and η_p, respectively.

[1] In A and BM, a distinction is made between K, the component of total angular momentum along the symmetry axis, and Ω, the sum of the Ω_p for the binding states of the individual particles. However, for the ground state of an approximately axially symmetric nucleus, $K = \Omega$, and the distinction between K and Ω is of significance only for the structure of the vibrational spectrum (cf. BM, § VI c.iv).

The proper choice of $\Theta_i(\vec{x}_p)$, for a given system, is the one which minimizes the effect of the coupling term in (VIII) and so leads to solutions which have approximately the form

$$\Psi = \Phi_\nu(\Theta_i)\,\psi_n(q_\alpha), \tag{XI}$$

where ν represents the rotational and n the intrinsic quantum numbers. Such a separation of the motion is expected to be possible if the rotational frequencies are small compared to the basic frequencies characterizing the internal motion. For a non-rigid system, where the moment of inertia depends on the deformation, this frequency condition is essentially equivalent to the assumption that the density fluctuations are small compared to the average deformation (cf., A, § V. 4).

As we shall see in the following section, it is also possible indirectly to characterize the functions $\Theta_i(\vec{x}_p)$, which accomplish the separation of the motion, by the equations governing the collective flow in the rotating body. By means of these equations, the rotational properties can be related to the average density distribution without any explicit consideration of the interactions in the system.

c) The hydrodynamical equations.

The application of collective co-ordinates in the description of a many-particle system implies the introduction of a collective flow, which is the quantum analogue of the classical velocity field in a macroscopic body.

According to (V), the velocity operator of a particle contains a collective part (VII) which, provided the fluctuations in the density distribution are small, depends only on Θ_i and the position $\vec{x}_p$ of the particle in question, but is practically independent of the position of the other particles. Moreover, if $\Theta_i(\vec{x}_p)$ is symmetric in the $\vec{x}_p$, this collective velocity is the same for all particles passing a given point $\vec{x}$ in the system.

Under these circumstances, we have to do with a collective flow, which we may define by

$$\vec{v}_{\mathrm{col}}(\vec{x}) = \langle n \,|\, \vec{v}_p^{(c)}(\vec{x}_p = \vec{x}) \,|\, n \rangle, \tag{XII}$$

representing an average over the intrinsic motion characterized by the wave function $\psi_n(q_\alpha)$ (cf. (XI)).

For each point $\vec{x}$, the velocity field (XII) is a function of the quantum mechanical variables Θ_i and their canonical conjugates. It is noted that the co-ordinates of $\vec{x}$ are to be regarded as independent of Θ_i only if referring to a body-fixed co-ordinate frame (x', y', z'). Thus, the components of $\vec{x}$ in a space-fixed frame (x, y, z) are functions of Θ_i, given by relations similar to (I). This implies that the velocity (VII), as well as the flow (XII), are Hermitian operators only if considered in terms of their components in the rotating, body-fixed co-ordinate system. In the following, we therefore, for convenience, assume the vector equations governing the collective flow to be decomposed in this manner.

From (VII) and (XII) follows directly

$$\operatorname{rot} \vec{v}_{\mathrm{col}}(\vec{x}) = 0 . \tag{XIII}$$

This relation is independent of the special choice of $\Theta_i(\vec{x}_p)$ and is thus an immediate consequence of the symmetry of the collective co-ordinates, which is implied in (XII), and which is characteristic of a shell structure.

In the case of a rigid structure, for which the collective co-ordinates are given by the unsymmetric equations (III), the collective part (VII) of the velocity of a particle depends essentially on its equilibrium position. Thus, the definition (XII) would imply that each particle takes part in the collective motion with its own velocity field. To obtain a collective flow, which is a unique function of $\vec{x}$, one may instead use the definition

$$\vec{v}_{\mathrm{col}}(\vec{x}) = \langle n \,|\, \vec{v}_p^{(c)}(\vec{\xi}_p = \vec{x}) \,|\, n \rangle . \tag{XIV}$$

This velocity field is no longer irrotational, but satisfies the condition

$$\operatorname{rot} \vec{v}_{\mathrm{col}}(\vec{x}) = 2\vec{\omega} \tag{XV}$$

characteristic of a rigid rotation. The vector $\vec{\omega}\ (\Theta_i, \dot{\Theta}_i)$ is the operator, representing the angular velocity of the intrinsic co-ordinate frame with respect to the fixed frame.

The separation between intrinsic and rotational motion also makes possible the definition of a collective density by the equation

$$\varrho_{\text{col}}(\vec{x}) = \langle n | \varrho(\vec{x}) | n \rangle, \tag{XVI}$$

where

$$\varrho(\vec{x}) = \sum_p \delta(\vec{x} - \vec{x}_p) \tag{XVII}$$

is the particle density operator. The collective density, as a function of (x', y', z'), is independent of Θ_i and describes the shape of the system.

The collective flow and density satisfy a continuity equation, which is derived from the particle conservation law

$$\operatorname{div} \varrho\vec{v} + \dot{\varrho} = 0, \tag{XVIII}$$

where the current operator $\varrho\vec{v}$ is given by (cf. (V))

$$\varrho\vec{v} = \frac{1}{2} \sum_p \delta(\vec{x} - \vec{x}_p)(\vec{v}_p^{(r)} + \vec{v}_p^{(c)}) + \text{Herm. conj.}, \tag{XIX}$$

and where $\dot{\varrho}$ is the time derivative of (XVII) for constant x, y, z. If (XVIII) is averaged over the internal co-ordinates, only the collective parts of $\varrho\vec{v}$ and $\dot{\varrho}$ contribute, and one obtains

$$\left.\begin{aligned} \operatorname{div} \varrho_{\text{col}} \vec{v}_{\text{col}} &= -\dot{\varrho}_{\text{col}} \\ &= \omega_{z'} \left(x' \frac{\partial}{\partial y'} - y' \frac{\partial}{\partial x'} \right) \varrho_{\text{col}}(x', y', z') + \text{cycl. perm.}, \end{aligned}\right\} \tag{XX}$$

where $\omega_{z'}(\Theta_i, \dot{\Theta}_i)$ is the angular velocity with respect to the intrinsic z'-axis. The cyclic permutations give the contributions to $\dot{\varrho}_{\text{col}}$ from rotations about the x'- and y'-axes.

The two equations (XIII) and (XX) are the direct quantum mechanical analogues to the hydrodynamical equations for an irrotational fluid (cf. (6), p. 16). For a given density $\varrho_{\text{col}}(x', y', z')$, they determine the collective flow as a function of the angular velocity $\vec{\omega}$, just as in the classical case.

From this flow, the rotational energy can be obtained from the familiar relation

$$H_{\text{rot}} = \frac{1}{2} m \int \varrho_{\text{col}} (\vec{v}_{\text{col}})^2 \, d\tau, \tag{XXI}$$

where $d\tau$ is the volume element in $\vec{x}$-space. Since $\vec{v}_{\text{col}}$, according to (XX), is linear in $\vec{\omega}$, the energy (XXI) becomes a quadratic function in $\vec{\omega}$, the coefficients of which define the effective moments of inertia. The result is exactly the same as for the corresponding classical calculation (cf. § III).

As an illustration, we consider an especially simple solution of the hydrodynamical equations[1], which is obtained for density distributions of the form

$$\varrho_{\text{col}} = \varrho_{\text{col}}(\lambda) \quad \text{with} \quad \lambda = \frac{x'^2}{a^2} + \frac{y'^2}{b^2} + \frac{z'^2}{c^2}, \qquad \text{(XXII)}$$

where a, b, and c are constants. In this case, equation (XX) is satisfied by an incompressible flow

$$\operatorname{div} \vec{v}_{\text{col}} = 0, \qquad \text{(XXIII)}$$

so that in the body-fixed co-ordinate system the streamlines follow the ellipsoidal surfaces of constant density (cf. Fig. 2d).

In order to show this and to determine $\vec{v}_{\text{col}}$, we introduce the velocity potential χ by (cf. (XIII))

$$\vec{v}_{\text{col}} = -\vec{\nabla}\chi. \qquad \text{(XXIV)}$$

Equation (XXIII) then implies

$$\Delta\chi = 0, \qquad \text{(XXV)}$$

or that χ is a linear combination of solid harmonics. By means of (XXII), (XXIII), and (XXIV), the continuity equation (XX) reduces to

$$\vec{\nabla}\lambda \cdot \vec{\nabla}\chi = -\omega_{z'}\left(x'\frac{\partial}{\partial y'} - y'\frac{\partial}{\partial x'}\right)\lambda + \text{cycl. perm.}, \qquad \text{(XXVI)}$$

which is satisfied by the velocity potential

$$\chi = -\frac{a^2 - b^2}{a^2 + b^2}\, x' y' \omega_{z'} + \text{cycl. perm.}. \qquad \text{(XXVII)}$$

Since this expression also satisfies (XXV), we have obtained the solution to the basic equations (XIII) and (XX).

[1] This solution has been given by A. R. Edmonds and M. Trocheris (private communications). It comprises, as a special case, the ellipsoidal nucleus of constant density, considered in A (§ II.3) for small values of the eccentricities.

From (XXVII) and (XXI) follows

$$H_{\text{rot}} = \frac{1}{2} m \left(\frac{a^2 - b^2}{a^2 + b^2}\right)^2 \int (x'^2 + y'^2) \varrho_{\text{col}} \, d\tau \, \omega_{z'}^2 + \text{cycl. perm.}, \quad \text{(XXVIII)}$$

and the moments of inertia are thus given by

$$\mathfrak{J}_{z'} = m \left(\frac{a^2 - b^2}{a^2 + b^2}\right)^2 \int (x'^2 + y'^2) \varrho_{\text{col}} \, d\tau = \frac{m \left(\int (x'^2 - y'^2) \varrho_{\text{col}} \, d\tau\right)^2}{\int (x'^2 + y'^2) \varrho_{\text{col}} \, d\tau} \quad \text{(XXIX)}$$

and cyclic permutations. The same result may be obtained by considering the rotational angular momentum

$$\vec{\mathfrak{P}} = m \int (\vec{x} \times \vec{v}_{\text{col}}) \varrho_{\text{col}} \, d\tau \quad \text{(XXX)}$$

and evaluating the components along the intrinsic axes.

It is seen that (XXIX) is equivalent to (IX) and, indeed, it can be shown that the collective flow (XXVII) is just that obtained, by means of (XII), from the definition (II) of $\Theta_i(\vec{x}_p)$. This choice of the collective co-ordinates is thus appropriate to density distributions of the type (XXII).

For a rigid rotation, the equations (XV) and (XX) have the solution

$$\vec{v}_{\text{col}} = \vec{\omega} \times \vec{x}, \quad \text{(XXXI)}$$

irrespective of the density distribution. This result is also obtained from (XIV), by using the definition (III) for $\Theta_i(\vec{x}_p)$.

While the discussion in this Appendix has been confined to to the especially simple collective properties of a many-particle system, associated with a pure rotational motion, the treatment can be directly extended to more general collective types of motion, involving oscillations in shape and density distribution. One must then introduce additional collective co-ordinates to describe these degrees of freedom, and the collective flow becomes a function also of the corresponding dynamical variables (cf. A; BM).

Dansk Resumé.

Atomkernernes Rotationstilstande.

Medens protonerne og neutronerne i en atomkerne er organiseret i en såkaldt skalstruktur, der minder om opbygningen af atomernes elektronsystemer, udviser kernerne også, i modsætning til atomerne, kollektive bevægelsesformer, hvori systemet som helhed deltager. Kernernes energispektre indeholder således to anslagstyper, den ene svarende til ændringer af de enkelte partiklers bindingstilstande i kernefeltet, den anden svarende til kollektive svingninger (Kapitel I).

En særlig simpel form for kollektive anslagstilstande optræder for stærkt deformerede kerner og svarer til en rotation af kernens orientering uden ændring af dens form eller indre struktur. Rotationsbevægelsen er væsensforskellig fra et stift legemes og kan betragtes som en bølge, der forplanter sig langs kernens overflade (cf. Fig. 2). Medens kernens rotationsspektrum har samme form som for et stift legemes rotation (cf. (15)), er det effektive inertimoment mindre og afhænger af kernens deformation (Kapitel III). Den kollektive rotationsbevægelse kan beskrives ved et hastighedsfelt, der tilfredsstiller hydrodynamiske ligninger (Appendix).

Kernens ligevægtsform afhænger af protonernes og neutronernes konfigurationer. Hvis disse danner lukkede skaller, er formen kuglesymmetrisk, men jo flere partikler der tilføjes til sådanne afsluttede konfigurationer, jo større bliver kernes deformation (Kapitel II; cf. Fig. 1). Med voksende deformation øges inertimomentet, og kvanteenergien for rotationsbevægelsen aftager (cf. Fig. 4).

Eksistensen af et rent rotationsspektrum kræver en så langsom rotationsbevægelse, at kernens form og indre struktur derved ikke influeres. Dette gælder for kerner med stor deformation, men med aftagende kernedeformation vil den øgede rotationshastighed perturbere kernens struktur og medføre afvigelser fra det simple rotationsspektrum. Det empiriske kendskab til kernernes lavt liggende anslagstilstande, der hidrører fra undersøgelser af finstrukturen i α- og β-sønderdelinger, af henfaldet af isomere tilstande, og af den ved Coulomb-excitation fremkaldte stråling, bekræfter disse forhold og viser, at rotationsspektrene optræder med særlig stor nøjagtighed for de stærkt deformerede kerner i massetalsområderne $155 < A < 185$ og $A > 225$ (Kapitel IV; cf. Fig. 5, 6, 7).

Den kollektive karakter af rotationsbevægelsen træder tydeligt frem i de målte elektromagnetiske overgangssandsynligheder mellem to rotationstilstande i samme kerne. Overgangssandsynlighederne, som bestemmes ved tilstandenes levetider eller ved virkningstværsnittet for Coulomb-excitation, er op til hundrede gange større end hvad der ville svare til kvanteovergange for en enkelt kernepartikel. Ud fra de målte overgangssandsynligheder kan kernedeformationen bestemmes (Kapitel V; cf. Tabel I).

References.

K. Alder and A. Winther (1953), Phys. Rev. **91**, 1578.
J. R. Arnold (1954), Phys. Rev. **93**, 743.
F. Asaro and I. Perlman (1952), Phys. Rev. **87**, 393.
F. Asaro and I. Perlman (1953), Phys. Rev. **91**, 763.
J. M. Blatt and V. F. Weisskopf (1952), Theoretical Nuclear Physics. J. Wiley and Sons, New York.
F. Bloch (1933), Zs. f. Phys. **81**, 363.
A. Bohr (1951), Phys. Rev. **81**, 134.
A. Bohr (1952), Dan. Mat. Fys. Medd. **26**, no. 14.
A. Bohr and B. R. Mottelson (1952), Physica **18**, 1066.
A. Bohr and B. R. Mottelson (1953), Dan. Mat. Fys. Medd. **27**, no. 16.
A. Bohr and B. R. Mottelson (1953a), Phys. Rev. **89**, 316.
A. Bohr and B. R. Mottelson (1953b), Phys. Rev. **90**, 717.
N. Bohr (1936), Nature **137**, 344.
N. Bohr and F. Kalckar (1937), Dan. Mat. Fys. Medd. **14**, no. 10.
B. M. Brown and D. H. Tomboulian (1952), Phys. Rev. **88**, 1158 and **91**, 1580.
R. S. Caird and A. C. G. Mitchell (1954), Phys. Rev. (in press).
E. L. Church and M. Goldhaber (1954), Bull. Am. Phys. Soc. (in press).
J. T. Eisinger, C. F. Cook, and C. M. Class (1954), Phys. Rev. (in press).
L. G. Elliot, R. L. Graham, J. Walker, and J. L. Wolfson (1954), Phys. Rev. **93**, 356.
K. W. Ford (1953), Phys. Rev. **90**, 29.
M. Goldhaber and R. D. Hill (1952), Rev. Mod. Phys. **24**, 179.
M. Goldhaber and A. W. Sunyar (1951), Phys. Rev. **83**, 906.
O. Haxel, J. H. D. Jensen, and H. E. Suess (1950), Zs. f. Physik **128**, 295.
O. Haxel, J. H. D. Jensen, and H. E. Suess (1952), Erg. d. Exakt. Naturw. **26**, 244.
A. Hedgran and D. Lind (1952), Arkiv för Fysik **5**, 177.
C. T. Hibdon and C. O. Muehlhause (1952), Phys. Rev. **88**, 943.
D. L. Hill and J. A. Wheeler (1953), Phys. Rev. **89**, 1102.
R. W. Hoff (1953), UCRL-2325, Berkeley, California.
J. M. Hollander, I. Perlman, and G. T. Seaborg (1953), Rev. Mod. Phys. **25**, 469.

T. Huus and J. Bjerregård (1953), Phys. Rev. **92**, 1579.
T. Huus and Č. Zupančič (1953), Dan. Mat. Fys. Medd. **28**, no. 1.
M. H. Johnson and E. Teller (1954), Phys. Rev. **93**, 357.
H. B. Keller and J. M. Cork (1951), Phys. Rev. **84**, 1079.
H. Kopfermann (1940), Kernmomente. Akad. Verl.-Ges. m. b. H., Leipzig.
H. Lew (1953), Phys. Rev. **91**, 619.
J. Lindhard and M. Scharff (1953), Dan. Mat. Fys. Medd. **27**, no. 15.
J. E. Mack (1950), Rev. Mod. Phys. **22**, 64.
M. G. Mayer (1950), Phys. Rev. **78**, 16.
C. L. McClelland and C. Goodman (1953), Phys. Rev. **91**, 760.
C. L. McClelland, H. Mark, and C. Goodman (1954), Phys. Rev. **93**, 904.
F. K. McGowan (1954), Phys. Rev. **93**, 471.
J. W. Mihelich, G. Scharff-Goldhaber, and M. McKeown (1954), Bull. Am. Phys. Soc. **29**, no. 1 (Y 5).
J. O. Newton and B. Rose (1954), Phil. Mag. **45**, 58.
H. H. Nielsen (1951), Rev. Mod. Phys. **23**, 90.
D. Pfirsch (1952), Zs. f. Physik **132**, 409.
J. Rainwater (1950), Phys. Rev. **79**, 432.
F. Rasetti and E. C. Booth (1953), Phys. Rev. **91**, 315.
J. O. Rasmussen (1954), Arkiv för Fysik, **7**, 185.
M. E. Rose, G. H. Goertzel, B. I. Spinrad, J. Harr, and P. Strong (1951), Phys. Rev. **83**, 79.
S. Rosenblum and M. Valadares (1952), C. R. **235**, 711.
G. Scharff-Goldhaber (1953), Phys. Rev. **90**, 587.
T. Schmidt (1943), Zs. f. Physik **121**, 63.
P. Stähelin and P. Preiswerk (1953), Nuovo Cimento **10**, 1219.
G. M. Temmer and N. P. Heydenburg (1954), Phys. Rev. **93**, 351 and 906.
K. A. Ter-Martirosyan (1952), Zh. Eksper. Teor. Fiz. **22**, 284.
R. van Wageningen (1953), Physica **19**, 1004.
R. van Wageningen and J. de Boer (1952), Physica **18**, 369.
L. Wilets (1954), Dan. Mat. Fys. Medd. (in press).

COMPTES RENDUS

DU

CONGRÈS INTERNATIONAL DE PHYSIQUE NUCLÉAIRE

30e anniversaire de la découverte de la radioactivité artificielle

PARIS - PALAIS DE L'UNESCO, 2-8 JUILLET 1964

préparés par Madame P. GUGENBERGER

Volume I

Rapports et discussions

(Extrait)

ÉDITIONS DU CENTRE NATIONAL DE LA RECHERCHE SCIENTIFIQUE
15, quai Anatole-France - PARIS-VII
1964

ELEMENTARY MODES OF NUCLEAR EXCITATIONS AND THEIR COUPLING

A. BOHR
Institute for Theoretical Physics
University of Copenhagen

When the Conference Secretary asked me for a contribution to the open session, I was very hesitant, having nothing new to report that has not been discussed at the other sessions. All I could do would be to make some general remarks in the form of a sketchy survey of some of the problems of nuclear spectra. This will at least give me the opportunity to mention a number of open problems.

Elementary modes of excitation

Let me first remind you that it has been possible to interpret a very extensive body of evidence on the nuclear spectra in terms of quite a simple picture, involving just a few types of excitations, the elementary modes of nuclear excitation. These may be grouped into three classes : *particles* (or quasi-particles), *vibrations,* and *rotations.*

The kernel of the particle excitations is nucleons or nucleon holes moving in the average nuclear potential. However, the actual excitons are affected to an important extent by numerous polarization and interaction effects which renormalize the charge and moments of the particle.

The kernel of the vibrational modes is a correlated two-body system, a particle-hole pair, or two particles (or two holes).

One of the challenging current problems is the search for new vibrational modes. A great richness is expected, corresponding to the, many quantum numbers which characterize the vibrational modes.

For a spherical equilibrium, the modes can, of course, be characterized by the angular momentum (λ, μ) and the parity π. Further we have the isospin quantum numbers τ, τ_z which express the isospin structure of the operator creating the mode in question. We can have $\tau = 0$ or 1, depending on whether the neutrons and protons are moving in phase or in opposite phase. Thus, the low lying quandrupole and octupole modes which are related to surface vibrations have $\tau = 0$, while the dipole photoresonance has $\tau = 1$.

In light nuclei, the ground state with no excitons (the "vacuum" state) is invariant with respect to rotations in isospace ($T = 0$) and the modes with different τ_z are simply related. In heavy nuclei, where the Coulomb field produces an alignment of isospins, one can still attempt to characterize the excitations by their τ-value, but the modes with different τ_z may have entirely different properties. Among the modes with $\tau_z = +1$ are the $\lambda = 0+$ excitations of the isobaric analogue state, which is now receiving much attention. In addition, however, one expects charge exchange modes which are not simply analogue states of modes with $\tau_z = 0$.

Next, we have the spin quantum number $\sigma = 0$ or 1, indicating whether nucleons with spin up and down move in phase or in opposite phase. The modes just mentioned seem to have $\sigma \approx 0$; little is as yet known about the excepted modes with $\sigma \approx 1$. The spin-orbit splitting which implies that the "vacuum" state may have unsaturated spins gives a new dimension to the σ-structure of the collective modes.

Finally, we can assign a particle (or baryon) quantum number β. For the particle excitons, we have $\beta = \pm 1$, for a particle or hole. For the bosons, $\beta = 0$ for a mode associated with oscillations in the average field, or a particle-hole mode, while $\beta = +2$ (or -2) for an exciton which involves the addition (or subtraction) of two nucleons to the nucleus. The " excitation " which takes us from the ground state of an even nucleus to the ground state of the next even nucleus is of this type (pairing mode with $\lambda\pi = 0+$). Other pairing modes have so far received only little attention. The structure of the excitations with $|\beta| > 2$ expresses the various possible clustering effects of corresponding order.

The quantum numbers τ, σ, and β are not exact; on account of the Coulomb, spin orbit, and pairing effects the modes may have mixed character. When one further notes that, because of the shell structure, there may be several distinct modes with each set of symmetry quantum numbers, one sees that we are in a virtually unexplored field.

The symmetry quantum numbers can be assigned on the basis of the selection rules governing the excitations in the different types of reactions. For example, $\beta = 0$ modes are strongly excited in inelastic reactions. More specifically (α, α') excite only modes with $\tau = 0$ and $\sigma = 0$. Modes with $\sigma = 1$ can be excited, for instance, by nucleon or electron scattering, where they reveal themselves by spin flip effects, and $\tau = 1$ modes can be excited by charge exchange effects ($\tau_z = \pm 1$) or by nucleon scattering ($\tau_z = 0$).

The $\beta = \pm 1$ modes are, as is well known, populated in one-particle transfer reactions (stripping, etc.). The $|\beta| = 2$ modes may be especially studied by two-nucleon transfer reactions, such as (p, t), etc.

While quantum number assignment can already be made on the basis of a rather qualitative analysis of the reactions involved, considerable progress is being made in the quantitative determination of the characteristic excitation matrix elements on the basis of the observed cross-sections.

In recent years, quite a considerable effort has gone into the calculation of excitation modes on the basis of the microscopic theory. Quantitative results are difficult to obtain, on account of uncertainties in the parameters involved associated with single-particle properties as well as with the very delicate problem of the effective interactions. However, even simplified calculations have proved very valuable in exploring the type of correlations which may occur, in suggesting significant new experiments, and in helping to see when the experimental results are really surprising.

Coupling between modes.

As the next step, we consider the coupling between the elementary modes of excitation, a problem intimately connected with the internal structure of the excitons themselves.

Coupling between rotation and intrinsic motion (particle and vibrational)

This coupling can be expressed in terms of Coriolis and centrifugal effects, and on account of the slowness of the rotational motion the coupling is relatively weak, i.e. the coupling energy is small compared to intrinsic excitation energies. This makes it possible to treat the coupling by a perturbation treatment which gives the nuclear properties as an expansion in I. In favourable cases, the expansion parameter may be of order (10^{-2} to 10^{-3}) $I(I+1)$, and, hence, quite small for not too large I.

The spectra of the strongly deformed nuclei are therefore amenable to analysis of an accuracy exceptional in nuclear physics, and are yielding a great deal of detailed insight into a variety of nuclear properties. What happens as I increases ? Many interesting phenomena may occur. Thus, one expects a gradual, or perhaps rather sudden breakdown of the pairing which will be very instructive to follow in detail. In addition, there are centrifugal distortions which, for sufficiently high I, may produce major changes in shape and even make contact with fission. In light nuclei, one is faced with the problem of whether the bands break off, as they will if the particles remain in the lowest major shells. However, we are learning that often it does not take so much energy to lift pairs of particles into higher shells.

So the surge towards higher angular momenta, with the aid of heavy ions, is one of the promising expansions of the nuclear horizon. In addition to the rotational angular momenta of the type considered, associated with shape deformations, we also have the possibility of collective angular momenta associated with the alignment of the angular momenta of the individual particles.

VIBRATION - VIBRATION COUPLING.

We now turn to the coupling between vibrational quanta. A problem of long standing and still unsolved is the anharmonicity in the low energy quadrupole spectra. It is evident that the effect is often very appreciable, but the situation varies from region to region, and the overall picture is far from clear. Let me mention two types of non-linearities. The very notion of independent vibrational excitations involves a semi-classical element, in the sense that it assumes a large number of particle degrees of freedom involved in the construction of the quanta. Possibly the finiteness of the degrees of freedom is essential for an improved description, so that one needs an even deeper contact than we have already with the shell-model approach, which involves an exact diagonalization of interacting particles, but in limited configurations. With the increasing power of computers and also through the aid of group-theoretical methods, one is now able to push this approach quite far, and this may bring a further elucidation of the structure of collective modes. It may also be that many of the nuclei usually considered as spherical should actually be regarded as transitional cases between spherical and deformed and thus require a more extensive analysis of the potential energy surface than has been attempted, so far. Indeed, the problem of further structure in the potential energy surfaces, and also of excited states with different shapes, looks very intriguing.

As an example of couplings between different types of vibrations, let us consider the coupling between the dipole photo-excitation and the low energy quadrupole modes. This coupling simply expresses the fact that the dipole frequency depends on the shape of the nucleus. The effect is well known to split the resonance in the deformed nuclei; also in nuclei with spherical equilibrium, the coupling may be strong, i.e. the coupling energy may be large compared to the quadrupole phonon energy. This poses a very interesting new type of problem, somewhat similar to the vibronic effect which the molecular physicists are engaged in tackling.

The dipole-quadrupole coupling has been studied in recent years by several investigators. It gives an effective broadening of the dipole resonance, and it is interesting that the line shape, or strength function, is not Lorentzian, but rather tends to be Gaussian, corresponding to the structure of the zero-point shape oscillations. This coupling appears, in many nuclei, to contribute a significant part of the observed line width, and with improved experimental techniques, quite a rich pattern of phenomena may become accessible. The whole problem of line shape and fine structure, and of damping phenomena, is indeed one about which we may learn a good deal in the coming years.

Coupling of particle to vibration.

What is the main coupling of particle motion to the internal nuclear degrees of freedom ? For small momentum transfers, one expects the main coupling to be associated with the excitation of vibrational modes. That is indeed what we see in direct inelastic scattering processes. The coupling simply arises from the fact that vibrations are associated with oscillations in the average nuclear field which act on the particle motion.

One has the opportunity to study this phenomenon in considerable detail in the spectra of odd-A nuclei. The coupling to the quadrupole excitations is an old problem of nuclear spectroscopy, which is only now beginning to yield to the combined probing with a multitude of agents, including particle transfer and inelastic scatterings, which together promise to make it possible to map in some detail the wave functions for the low energy states. One can get quite a variety of patterns representing the various stages from weak coupling to strong coupling involving a transition to rotational spectra. Usually, it seems that the coupling is of intermediate strength. It is somewhat weaker in deformed nuclei which have a more well-defined shape, but is also here beginning to manifest itself.

We are also beginning to obtain information on the coupling of particle to octupole vibrations. Here the coupling is often much weaker, which seems to be connected with the fact that the coupling necessarily involves a change of orbit of the particle, and that frequently no low lying orbits are available to which the particle can couple strongly. However, there are also signs of strong octupole distortions in certain configurations.

If we believe the stated premises concerning the coupling of particles to internal modes, we are similarly led to expect that the main coupling of an incident low energy particle is through the excitation of vibrational modes. This is a view rather different from that of the classical optical model. We here enter upon the whole problem of intermediate structure in the reaction processes which is now receiving much attention and which represents another possibility for probing deeper into the nuclear structure.

Particle-particle coupling.

Finally, we come to the particle-particle coupling.

The nucleonic forces are, of course, the source of the entire nuclear dynamics, but the idea has been to try to push ahead, without having to face directly the full complexity of these forces, and instead to absorb their main effects into the dynamical elements already considered. Thus, as a first step, we introduce the static nuclear field and the average pair correlations which define the particle excitons; secondly, we consider the dynamic aspects of the field and of the pairing, represented by the collective modes, and thirdly, the various couplings we have talked about, which are directly connected with the simple picture of the vibrational and rotational modes.

How far can we go in this direction, what couplings between particles remain ? We may learn a good deal about this problem when we obtain more detailed information about two-particle states in the nuclear spectra and see to what extent the main interaction effects can be absorbed into well-defined collective modes, and to what extent we shall be dealing with intermediate situations between two-particle states and collective modes.

For higher energies and momentum tranfers, we have to face more directly the particle-particle interaction, and, at the same time, the analysis of the quasi-particles, and other excitons in terms of the primordial nucleons.

In conclusion, I want to apologize for having said so much which is familiar to most of you and for talking in such general terms. I am also very much aware of the fact that, in trying to sketch broad features, I have been very schematical and have oversimplified many issues.

M. le Professeur Teillac. — Je remercie le Professeur Bohr de son intéressant aperçu général et je demande s'il y a des interventions sur cette présentation. *(Il n'y a pas de questions sur l'exposé de M. Bohr.)* Je remercie à nouveau le Professeur Bohr et je donne la parole au Professeur Van Hove.

Bohr arriving at the Institute on his bicycle, ca. 1969.

Reprint from

"CONTEMPORARY PHYSICS"

VOL. II

INTERNATIONAL ATOMIC ENERGY AGENCY
VIENNA, 1969

EXCITATIONS IN NUCLEI

A. BOHR
The Niels Bohr Institute, University of Copenhagen,
Copenhagen, Denmark

Abstract

EXCITATIONS IN NUCLEI. 1. Spectrum of elementary excitations; 2. Static deformations. Rotational spectra; 3. Relationship between members of a collective family; 4. Analysis of collective parameters; 5. Spurious degrees of freedom; 6. Pairing modes.

The nucleus consists of a handful of particles with fairly well established interactions, and so one might think that the only proper approach to the subject would be to start from the Schrödinger equation for the A nucleons and proceed by developing appropriate and well-defined approximation methods. However, even systems of few particles and a hundred turns out to be quite a handful, exhibit an enormously rich variety of structural facets and, as in other systems where many degrees of freedom are involved, the central problem becomes that of establishing the relevant degrees of freedom and of developing the appropriate concepts that govern their interplay. Progress in this direction has been achieved by a combination of approaches, at many levels of phenomenology. This makes it difficult to give anything like a systematic presentation of the subject, and we have selected a few problems of a fairly general nature that have puzzled nuclear physicists, and that might be suited to elicit response from the broader community of quantal physicists.

1. SPECTRUM OF ELEMENTARY EXCITATIONS

To introduce the quanta of nuclear physics, let us consider, as an example, the spectrum of elementary excitations based on the ground state of ^{208}Pb. This "vacuum" state has especially simple properties (has a minimum of degeneracy) on account of the closed-shell configuration of 82 protons and 126 neutrons.

The elementary modes of excitation of fermion type are shown in Fig. 1. The quantum numbers and approximate energies of the excitations within the first few MeV can be simply interpreted on the basis of single-particle motion in a central potential with spin-orbit coupling. This interpretation of the excitations is strikingly borne out by the single-nucleon transfer reactions (stripping and pick-up reactions, such as (d, p), (d, t), (t, α) etc.), which directly measure the matrix elements of the nucleon creation and destruction operators (the nucleon field operators), and which are found to populate just these levels with large cross-sections.

Other properties of the quasi-particles (single-particle and single-hole excitations), such as electromagnetic and β-decay transition moments, confirm the interpretation, but at the same time show renormalization effects due to the coupling to the underlying closed-shell structure. In the present example, the couplings are weak in the sense that each individual particle is only little perturbed, but the renormalization effects may become large due to the accumulative response of all the particles embedded in the vacuum. For example, a neutron carries an effective charge of about one unit, and β-decay moments are reduced, sometimes by an order of magnitude.

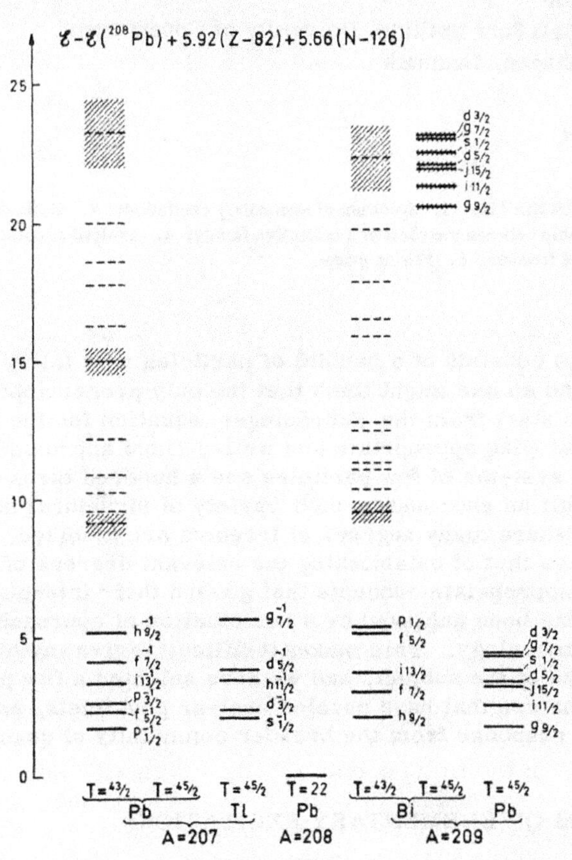

FIG. 1 Quasiparticle excitations based on ^{208}Pb. The dashed levels only illustrate schematically the spectrum at higher excitations.

The figure also indicates the pattern of higher-lying single-particle and single-hole states, which are still only very incompletely explored. Quite new problems arise because these states occur at excitations where the total level density is very high. Some of these problems will be taken up at a later session and we shall only mention the basic concept of strength function, which refers to the manner in which the properties of the elementary mode may become shared by a large number of neighbouring levels. The strength function is a generalization of the line structure of a decaying particle, and incorporates the consequences of the finiteness of the nuclear system and the associated lack of irreversibility of the decay process. A remarkable feature is the occurrence of quite sharp levels (the isobaric analogue states), which owe their stability to the approximate validity of the isospin quantum number, which persists in spite of the strength of the symmetry-violating Coulomb forces responsible for the large splitting between members of a T-multiplet.

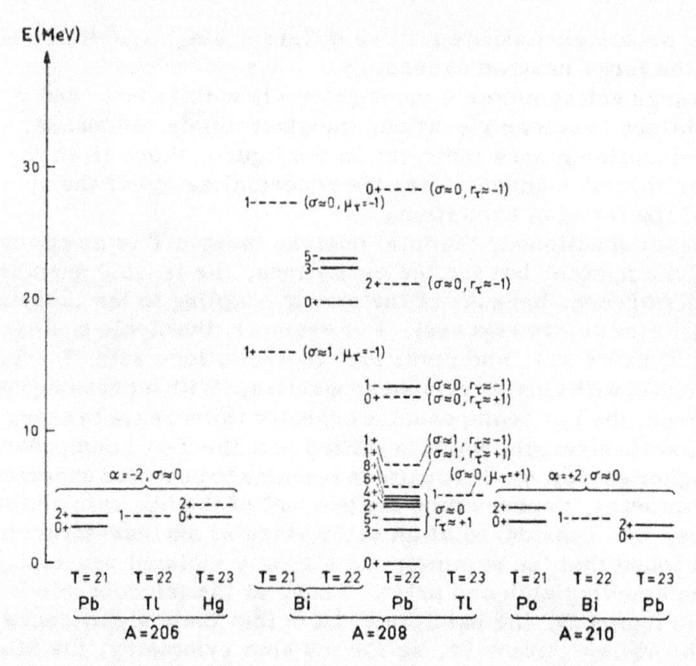

FIG. 2 Boson excitations based on ²⁰⁸Pb. The dashed levels represent theoretical estimates.

The boson excitations in ^{208}Pb are illustrated in Fig. 2, and show a rich spectrum, which is rapidly being supplemented by new discoveries. The excitations in the figure are collective in the sense that they have large matrix elements (in single-particle units), and thus involve many degrees of freedom of the nucleonic motion. Additional excitations are observed, which can be approximately described in terms of two elementary fermion excitations; on account of the finite geometry, the distinction between collective and two-particle excitations is, of course, not a sharp one.

The collective bosons can be associated with different types of deformations in the nucleon density and in the average potential. For example, the lowest excited states in ^{208}Pb correspond to shape oscillations of octupole and diatriacontrapole type. In addition to the quantum numbers λ and π (angular momentum and parity), one can characterize the excitations by additional, approximate-symmetry quantum numbers. Thus, the octupole mode may be assigned the spin quantum number $\sigma \approx 0$, since it is found that the oscillating nuclear potential associated with this mode has only a weak spin dependence. This symmetry belongs naturally to the macroscopic picture of a shape oscillation, but is remarkable because the strong spin-orbit coupling, which decisively influences the nuclear shell structure, violates the invariance with respect to rotations in spin space. Thus, in the microscopic description of the octupole mode, in terms of superpositions of excitations of individual nucleons, the σ-symmetry is completely broken. It appears on the macroscopic level, when we consider the long wave-length components of the oscillations in the nucleon density and potential, involving the superposition of many individual particle excitations. In a similar manner, one can characterize the collective properties of the excitations by a charge symmetry quantum number, r_τ; the octupole mode involves approximately charge symmetric collective density variations ($r_\tau = +1$), though the individual neutron and proton excitations involve different single-particle orbits, on account of the large neutron excess.

The charge antisymmetric mode (r_τ = -1) with $\lambda\pi$ = 1^- and $\sigma \approx 0$ is the oldest established nuclear vibration, the giant dipole resonance. For the σ = 1 spin vibration modes indicated in the figure, there is so far only indirect experimental evidence from the renormalization of the spin magnetic moments of the fermion excitations.

As already mentioned, the total nuclear isospin T is an approximate constant of the motion, but for the excitations, the isospin quantum number τ may be badly broken, because of the strong coupling to the isospin T_0 of the vacuum state (the neutron excess). For example, the dipole mode in a nucleus with small T_0 has τ = 1, and gives rise to excitations with $T = T_0$, $T_0 \pm 1$ forming a triplet with closely related properties. With increasing neutron excess, however, the $T_0 + 1$ component is expected to decrease in energy and to gradually lose its strength, which is shifted on to the $T_0 - 1$ component, which moves to higher energies. This pattern remains to be tested experimentally.

SU_4 symmetry, representing the product of the SU_2 spin and isospin symmetries, was considered at an early stage of nuclear-structure physics. It has been found that the symmetry is strongly violated especially by the interactions coupling spin and orbit. Thus, at the microscopic level, the SU_4 quantum numbers, the partitions, have lost their significance except in certain light nuclei. However, as for the spin symmetry, the SU_4 symmetry may reappear at the macroscopic level. For example, the pattern of 0+ and 1+ collective modes with T = 22 and 21, respectively, are expected to have relations corresponding qualitatively to a super-multiplet of a high dimensionality ([f] = [22, 22, 0, 0]).

The modes with nucleon transfer number α = ± 2 may be viewed as oscillations in the nuclear pairing field, which represents the creation and annihilation of two nucleons at the same point in space or close together. The pairing effect, responsible for the difference in binding between even and odd nuclei, has been known from the early days of nuclear physics, but its many-particle structure was only understood after the advent of the microscopic theory of superconductivity. A basic manifestation of the correlation effect is that it gives rise to collective transitions between nuclei differring by two particles. These transitions are now being studied by reactions such as (t, p), (^{3}He, p), etc., by which two nucleons are simultaneously exchanged between projectile and target, and the α = ± 2 modes in Fig. 2 are found to be excited with strongly enhanced matrix elements.

2. STATIC DEFORMATIONS. ROTATIONAL SPECTRA

The boson spectrum considered above has a vibrational character, i.e. it corresponds to oscillations of the various density components about the equilibrium represented by the vacuum state (the ground state of ^{208}Pb). However, for nuclei with many particles in unfilled shells, the independent - particle motion gives rise to a highly degenerate ground state, and the system lifts the degeneracy by reducing the symmetry, i.e. by deforming itself (a well-known feature of quantal systems, referred to in molecules as the Jahn-Teller effect). In nuclei, the most important stable deformations occur in the $\lambda\pi = 2^+$, α = 0 channel (quadrupole shape deformations) and in the $\lambda\pi = 0^+$, α = ± 2 channel (monopole pairing deformation (superfluidity).

Stable deformations lead to the occurrence of families of collective excitations with a rotational rather than vibrational relationship, and with degrees of freedom depending on the symmetry violation in the intrinsic motion caused by the deformation. Some examples will illustrate the variety of patterns that may arise.

Let us first consider the breaking of invariance with respect to rotations $\mathscr{R}$ in ordinary three-dimensional space. The separation between collective and intrinsic motion corresponds to a Hamiltonian of the form

$$H = H_{intr} + H_{coll} \tag{1}$$

where H_{coll} depends on the collective quantum numbers, such as the angular momentum I while H_{intr} describes motion relative to the intrinsic frame specified by the collective orientation angles, the conjugates of the angular momentum variables. Thus, the total wave function has the form

$$\Psi = \Phi_{int}(q)\, \mathscr{D}_{coll}(\theta) \tag{2}$$

where q are intrinsic variables (rotational scalars) while θ represents the collective angles. In the extreme case of symmetry violation, the intrinsic deformed system does not go into itself by any rotation of the co-ordinate system. The family of collective states then has the full degrees of freedom of the asymmetric rotor, i.e. there are three collective orientation angles (the Euler angles) and the angular momentum variables I, M, and K (the projection of I on an intrinsic axis), of which K will not in general be a constant of the motion. Thus, the collective spectrum comprises $(2I + 1)$ states for each value of I and M, and therefore forms a two-dimensional system of trajectories in angular momentum space. This situation occurs in molecules, but in nuclei the observed deformations preserve part of the rotational invariance, with a resulting reduction in the collective degrees of freedom. Thus, if the intrinsic Hamiltonian is invariant with respect to a rotation $\mathscr{R}$, this operation is an instrinsic degree of freedom, $\mathscr{R}_i$, and the collective motion must satisfy the constraint that $\mathscr{R}_e$ (the rotation $\mathscr{R}$ considered as an external operation acting on the orientation angles) equals $\mathscr{R}_i$. For example, the well-established nuclear deformations have axial symmetry; the component K of angular momentum with respect to the symmetry axis is therefore an intrinsic property, and the rotational band is reduced to a single trajectory with $I = K, K + 1, \cdots$. A further reduction of the rotational degrees of freedom occurs if H_{int} is invariant with respect to finite rotations, such as a rotation $\mathscr{R}_2(\pi)$ of π about an axis (2) perpendicular to the symmetry axis. For $K = 0$, we then have bands with $I = 0, 2, \cdots$, well known for the ground-state configurations of even-even nuclei.

New patterns arise if the deformations are linked to the violation of other symmetries; thus, a breaking of $\mathscr{P}$-symmetry (space reflection) gives rise to parity doublets, but if the deformation retains the combined symmetry $\mathscr{R}_2(\pi) \cdot \mathscr{P}$, we obtain, for $K = 0$, a band with $I = 0^+, 1^-, 2^+ \cdots$. A link between rotational and isospin symmetry breaking is, as is well known, a possible feature in the structure of the nucleon itself. For example, in the strong coupling model of nucleon-pion interaction, the deformed meson field is invariant with respect to simultaneous rotations in spin-orbital and isospin space and gives rise to a rotational band with $I = T$.

The pairing deformations provide a new dimension to the collective families. These deformations violate the gauge invariance associated with nucleon number conservation, and give collective branches with baryon numbers A, A ± 2, A ± 4, ··· representing rotational motion in gauge space. The deformations conserve the distinction between even and odd nuclei (are invariant with respect to gauge transformations $\mathscr{G}(\pi)$).

One notes the close similarity to the description of the superfluid system, which Anderson has discussed; the state with finite expectation value of the pairing field is here the intrinsic state, and the phase ϕ is the collective coordinate, corresponding to the orientation of a deformed system. In the nucleus, the main effects are associated with the part of the phase field that is a constant over the nucleus. The overall number conservation is explicit in the formalism; the symmetry is broken at the intrinsic level, and the pairing field is expressed in terms of intrinsic variables

$$\psi_{\mathrm{pairing}} \sim a^{\dagger}(r, m_s)\, a^{\dagger}(\overline{r, m_s})\ \exp\{-2i\,\phi_{\mathrm{m}_{\mathrm{coll}}}\} \qquad (3)$$

which commutes with the number operator. The direct manifestation of the coherence is in the collective "rotational" spectrum, the numbers of which are connected with large matrix elements for two-nucleon transfer.

A richer spectrum of pair addition modes occurs, when we include pair correlations between neutrons and protons. The main pairing effect occurs in the ^{1}S channel with isospin $\tau = 1$, and the pairing field therefore involves a deformation in isospace (the gap parameter Δ is an isovector). The favoured pairing field has axial symmetry and invariance with respect to the rotation $\mathscr{R}_2(\pi)$ in isospace multiplied by a gauge transformation $\mathscr{G}(\pi/2)$, and the collective rotational spectrum, for even A and $K_\tau = 0$, contains the states with even values of $T + A/2$. The ground states of the even-even nuclei, together with their isobaric analogue states, form a family of this type. (The problem is related to superfluidity associated with a condensate in states with finite angular momentum in the relative motion of the paired particles).

Another manifestation of the permanent distortions is in terms of the modified particle degrees of freedom. As already mentioned, the quasiparticle excitations can be directly studied in the one-particle transfer reactions. For example, if one considers an even-even deformed nucleus and transfers a particle into a given orbit ν in the deformed field, the intensity pattern of the rotational band associated with this orbit gives a fingerprint of the orbit, from which one can deduce the probabilities $c_\nu^2(j)$ that the particle possesses an angular momentum j, while the relative intensities for producing the states by adding a particle or removing a particle give the probability $v^2(\nu)$ that the single-particle orbit is occupied in the pair correlated phase.

3. RELATIONSHIP BETWEEN MEMBERS OF A COLLECTIVE FAMILY

The relationship between members of a collective family is expressed by the dependence of the different matrix elements on the collective quantum numbers. The problem has been especially studied for the rotational bands in I-space, where the empirical information is very rich. For not too large values of the angular momentum I the various matrix elements can be expressed in a power series in I. The form of this series follows from the

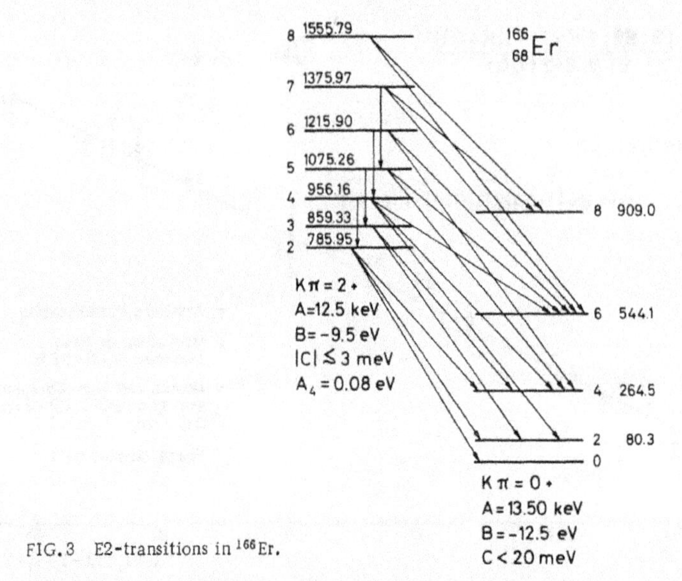

FIG. 3 E2-transitions in ^{166}Er.

symmetry of the intrinsic Hamiltonian, which, as mentioned above, governs the degrees of freedom of the collective motion.

The level of sophistication reached by the experimental studies may be illustrated by the E2-transitions between the $K = 2^+$ band and the $K = 0^+$ band of the nucleus ^{166}Er (see Fig. 3), which possesses a shape deformation with axial symmetry and $\mathscr{R}_2(\pi)$ invariance. The expansion of the E2-transition probability is obtained by transforming the E2-moment to the intrinsic frame with orientation ω,

$$\mathscr{M}(\mathrm{E2},\mu) = \sum_{\nu} \mathscr{M}'(\mathrm{E2},\nu)\, \mathscr{D}^2_{\mu\nu}(\omega) \tag{4}$$

The components $\mathscr{M}'(\mathrm{E2},\nu)$ are scalars and can be expanded in powers of the angular momentum components with respect to the intrinsic frame, and one obtains, for the reduced transition probability

$$\begin{aligned} B(\mathrm{E2};\, I_i K = 2 \rightarrow I_f K = 0) &= (2I_i + 1)^{-1} \langle I_f \| \mathscr{M}(\mathrm{E2}) \| I_i \rangle \\ &= \langle I_i\, 2, 2 - 2 | I_f 0 \rangle^2 (M_1 + M_2\,(I_i(I_i+1) - I_f(I_f+1)) + \ldots)^2 \end{aligned} \tag{5}$$

The experimental data appear to be well described by the first two terms in this expansion, corresponding to a straight line in Fig. 4. The leading-order term M_1 is related to the transition probability in a nucleus with fixed orientation, while M_2 expresses the perturbations caused by the Coriolis forces associated with the rotational motion. The establishment of such intensity relations provides a decisive test of the rotational coupling scheme. A large body of data on energies and matrix elements for weak, electromagnetic, and strong interaction transition moments is available and has been analysed to yield informations on the intrinsic structure and its coupling to the rotational motion.

For large quantum numbers I, the rotational perturbations may become so large that the power series expansion breaks down. In heavy nuclei, this occurs for $I \sim 10$-20 for the energy expansion. However, the family relationship may still be retained, manifesting itself in a smooth dependence of the

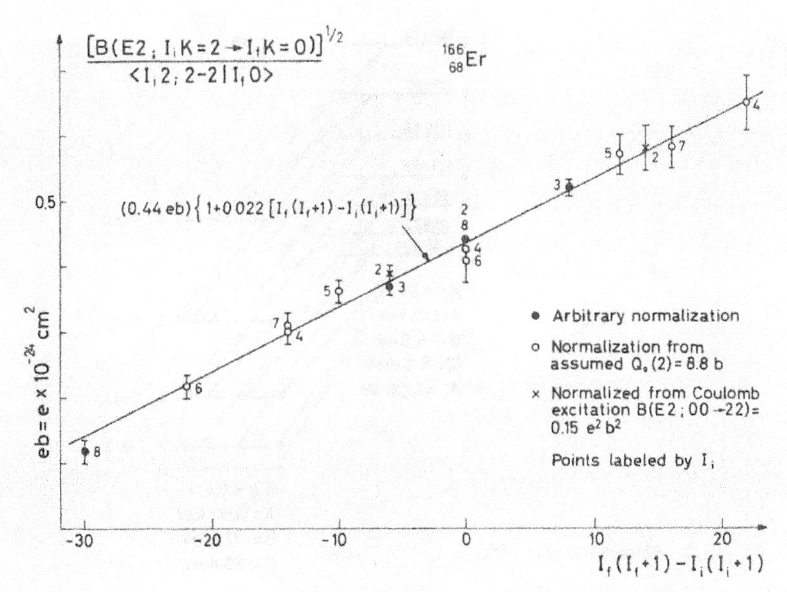

FIG. 4 Rotational intensity relation for E2-transitions in ^{166}Er.

properties on the collective quantum number. For example, for the collective family associated with the pair addition mode, the nuclear properties can be expanded in the collective quantum number A but not around A = 0, only in a local region around $A = A_0$.

Many still unanswered questions arise in connection with the possibility of discontinuities in the trajectories. Such discontinuities may occur as a result of the shell structure, i.e. of specific quantal features in the underlying intrinsic dynamics. For example, the pair addition mode is found to have a discontinuity at major shell closings and has been observed to fractionate at abrupt transitions between spherical and deformed nuclei.

Challenging problems are associated with the structure of rotational bands for very large values of I; a number of different effects may come into play, such as a transition from a superfluid to a normal Fermi system, major centrifugal distortions leading to asymmetries in the nuclear shape, as in the rotating stars, or to fission in the very heavy elements, and a discontinuity may occur when the angular momenta of the particles in the unfilled shells have achieved maximal alignment. In the heavy nuclei, it appears that the pairing is especially vulnerable to the effect of the rotation and, at the highest I-values studied, we are beginning to see effects indicative of the disappearance of the stable pairing field. The effect is quite similar to the disappearance of superconductivity at the critical magnetic field. The phenomenon may afford the opportunity for studying such a "phase transition" under favourable conditions due to the special simplicity of the pairing field, which, in first approximation, couples only the single-particle states that are conjugate under time reversal.

Recently, Anderson objected to talking about phase transitions in the small particle superconductors, and it is of course true that we do not need non-analytic functions to describe the transitions in the finite systems. On the other hand, the phenomena are closely related and, in a system like the nucleus, we have the possibility of studying the transition in terms of the spectrum of individual quantum states. Thus, the transition from pair-correlated to normal system with increasing angular momentum involves the coupling between the bands associated with the ground state and the excited bands representing fluctuations in the pairing field.

4. ANALYSIS OF COLLECTIVE PARAMETERS

As in other domains of quantal physics, the analysis of nuclear excitations involves a series of different levels of phenomenology. First, there is the general classification, in terms of symmetries, particle configurations, and collective families, on the basis of which the empirical data are expressed in terms of intrinsic parameters, such as the average nuclear potential, effective particle moments and interactions, parameters in the rotational expansions, or, in the case of vibrations, the restoring forces, mass parameters, anharmonicity parameters, etc.

On the next level, one may attempt to describe the collective nuclear properties in terms of macroscopic properties of nuclear matter, such as surface tension, compressibility, polarizability (or more general response functions) and the properties of the collective flow. Among the notable successes of this approach has been the determination of the eigenfrequency for the dipole mode on the basis of a two-fluid hydrodynamical model in which the coupling of the fluids can be related to the symmetry energy in the nuclear masses. However, such an approach is of limited scope owing to the important quantal effects associated with the shell structure, which deeply affect the low-energy collective nuclear properties. The effect of the shell structure on vibrational eigenfrequencies is illustrated in Fig. 5 and is seen to be especially pronounced for the quadrupole mode, where indeed it leads to instability of the spherical shape in certain regions of nuclei with a resulting splitting of all the modes.

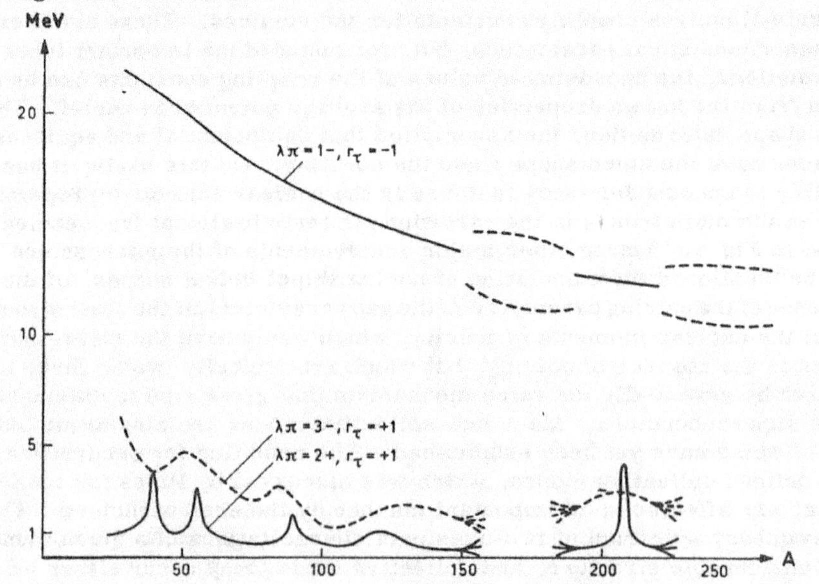

FIG. 5 Energies $\hbar\omega$ of the giant dipole resonance and the low-energy quadrupole and octupole shape oscillations.

The shell structure expresses the quantization of the nucleonic motion in the nucleus as a whole and therefore implies an essential non-locality in the collective nuclear properties. The inclusion of these effects involves going to a more microscopic level, which treats explicitly the degrees of freedom of the individual particles and the interactions which generate the correlated nucleonic motion. One may attempt directly to introduce the two-particle interactions, as determined from the two-nucleon scattering experiments, but the problem is a difficult one, because of the many correlation effects which must be considered; thus, the effective forces between nucleons in a nucleus are rather different from those of free nucleons.

In the analysis of the nuclear excitations, considerable success has been achieved by imposing another level of phenomenology, which can be formulated in various ways. We know from the existence of the nuclear shell structure that the interactions can be represented in first approximation by an average field, and, thus, we may attempt to describe the coupling between the particles and the collective motion in terms of oscillations in the nuclear field associated with collective deformations. In this manner, the interactions are described by a three-field coupling  representing the coupling of the particles to the nuclear deformation, rather than the four-field coupling  representing the nucleonic interactions. The approach is sometimes formulated in terms of simplified effective nucleonic interactions of separable type, such as the quadrupole force (representing the product of the quadrupole moments of the interacting particles) but these interactions are merely constructions designed so as to yield the three-field couplings, when treated in lowest order. The determination of the eigenfrequencies of the collective motion is then a self-consistency problem similar to that involved, for example, in the determination of the plasma frequency or zero sound, as described by Pines. In both cases, the interactions are introduced in terms of the relation between variations in density and potential.

This relation, between the nuclear deformations and the potentials they generate, involves coupling constants for the vertices. These are treated as phenomenological parameters, but, for some of the important types of deformations, the approximate values of the coupling constants can be deduced from the known properties of the average potential in nuclei. Thus, for a shape deformation, the assumption that equipotential and equidensity surfaces have the same shape fixes the coupling. On this basis, it has been possible to account for many features in the nuclear collective properties, such as the main trends in the variations in the vibrational frequencies illustrated in Fig. 5. Among other major achievements of the microscopic theory may be mentioned the calculation of nuclear equilibrium shapes, of the dependence of the pairing parameter Δ (the gap parameter) on the shell structure, and of the nuclear moments of inertia, which would have the rigid-body values in the absence of pairing, but which are typically two or three times smaller by essentially the same mechanism that gives rise to diamagnetism of the superconductors. Many new collective modes are also predicted, only part of which have yet been established. The condition for occurrence of well-defined collective modes, which was discussed by Pines for the infinite media, are affected in an important manner by the shell structure. Thus, the frequency spectrum of two-quasiparticle excitations of a given symmetry has considerable structure, and collective modes may occur either below the lowest intrinsic frequencies or above the highest, or whenever there is an appreciable gap in the intrinsic excitation spectrum.

In spite of the progress achieved, it must be emphasized that the microscopic analysis of the collective nuclear properties is still in a rather preliminary stage. It is important that the theory makes detailed predictions concerning the microscopic structure of the collective motion. With the tools available, this can be tested in great detail by a variety of reactions (see Fig. 6), such as (a) the production of phonons by transfer of a particle to a target with one quasi-particle, (b) the decay of phonons into two quasi-particles (if the phonons are energetically stable, one may consider the isobaric analogue states, which are unstable with respect to proton emission), (c) inelastic nucleonic scattering with the excitation of a phonon, (d) production of states with quasi-particle and a phonon, (e) energy splittings within the multiplet formed by a quasi-particle and a phonon, as well as a variety of higher-order effects, such as (f) the scattering of two phonons (anharmonicity in the vibrational motion).

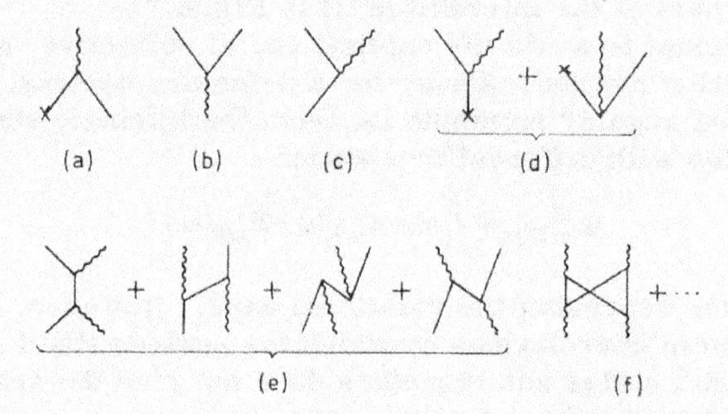

FIG. 6 Diagrams for various processes probing the structure of the phonons (see explanation in the text).

The field is a huge one and concerns the whole problem of the interaction between the elementary excitations, which is inseparable from the structure of the excitons themselves. It is an area of vigorous activity, and it is impossible on this occasion to mention the many new ideas and methods which are being pursued and explored. The perturbation expansion can only be directly applied for the modes involving a relatively weak coupling to the particle motion; however, in the case of special importance for the low-energy quadrupole mode, the coupling is strong or intermediate; the dimensionless coupling constant is for most nuclei greater than unity. In some situations, we have a genuine strong coupling phenomenon resulting in a stable deformation and associated rotational spectra, but for a large class of nuclei, we are dealing with a transitional type of spectra intermediate between rotational and harmonic vibrational. The strength of the coupling at the same time implies large anharmonicities in the vibrational spectra of even nuclei, i.e. interactions between the quanta comparable with their own frequency. These anharmonicity phenomena have for long presented a major challenge. Some insight has been gained by many different approaches, based on power series expansions of the Hamiltonian, analyses of the potential energy surfaces of deformation and of the collective flow, and attempts to isolate and specifically include the particle degrees of freedom mainly responsible for the anharmonicity. However, we are dealing with a delicate instability phenomenon, as can be seen from the fact that a quadrupole moment added to the system typically induces a moment 20-30 times larger. Peculiar phenomena may therefore occur and surprises may still be in stall.

5. SPURIOUS DEGREES OF FREEDOM

In the description of the nuclear dynamics in terms of collective and individual particle degrees of freedom, there arises the problem of the overcompleteness of the degrees of freedom, since the collective bosons are composites of the fermions. This is a general problem of many-body physics, but has been felt especially acute in the nucleus because of the finite geometry and the resulting discrete degrees of freedom. In a sense, the lesson has been that there is no cause for special worries, in that the elimination of the spurious degrees of freedom comes as part of the systematic analysis of the interaction effects. For example, the effect of the exclusion principle acting between a particle and a boson is contained in the interaction (e) in Fig. 6, and the fact that bosons consisting of fermions cannot be freely superposed, as a result of the exclusion principle for the fermions, is expressed as part of the interaction (f) in Fig. 6.

One may attempt to avoid the explicit use of collective variables by the so-called projection method. Thus, for a deformed nucleus, one may project states of specified angular momenta IM from the intrinsic state $\Phi_{\alpha,K}$, by superposing states with different orientation ω

$$\Psi_{\alpha,\,IKM} = \int d\omega\; \Phi_{\alpha,\,K}(\omega)\,\mathscr{D}^{I}_{MK}(\omega) \tag{6}$$

and in this manner represent the rotational band. However, this approach ignores the dynamic correlations by which the nucleus itself generates the family of rotational states and therefore does not give the separation of the motion expressed by the wave function (2).

6. PAIRING MODES

As a last topic, I should like to come back to the structure of the collective modes in the pairing channel. The development of this subject has been stimulated by recent studies of double transfer reactions.

We shall consider medium heavy nuclei, where the neutron excess is fairly small, and shall view the situation with the nucleus ^{56}Ni as a basis. This nucleus has N = Z = 28 and closed shells of neutrons and protons (like a solid with filled conduction bands). The excitations of the pairing field are therefore of vibrational type (Cooper pairs); there are two types of quanta corresponding to the addition of pairs into the shells above 28 ($\alpha = +2$, $\lambda\pi = 0^+$) and removal of pairs from the shells below 28 ($\alpha = -2$, $\lambda\pi = 0^+$): each pair has unit isospin, so in all there are six modes. The one-quantum excitations represent the ground states of ^{58}Ni and ^{54}Fe and their isobaric analogues. There is considerable coupling between the particles in the shells above and below 28, but this is included as a renormalization of the quanta. The excitations with two quanta are shown in Fig. 7. The full-drawn lines correspond to ground states, and the dashed ones to excited states, recently found in double-transfer reactions. On the basis of the observed interactions between the quanta, one can attempt to construct the pattern of states with more quanta, as is illustrated schematically in Fig. 8. Of the states with quanta of both types, only the few lowest are included in the figure.

This spectrum of excitations comprises all the ground states of even-even nuclei in this domain and, in addition, a host of other 0^+ states, which

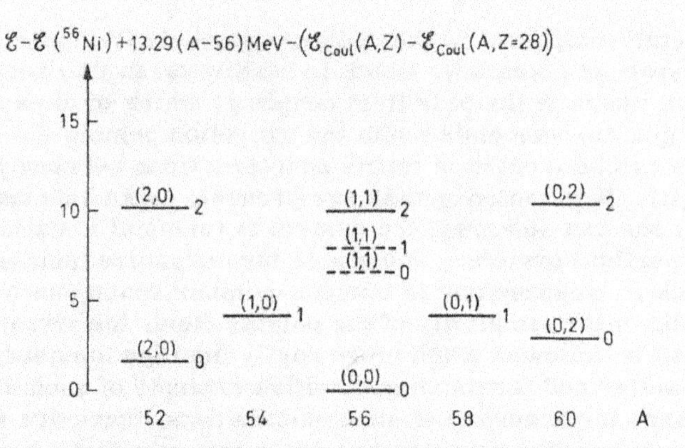

FIG.7 Pairing vibrational excitations with one and two quanta added to ^{56}Ni. The levels have $I\pi = 0^+$, and the quantum numbers (n_1, n_2) give the numbers of quanta with $\alpha = -2$ and $\alpha = +2$, respectively. The levels are also labelled by the total isospin T. Full-drawn lines represent ground states, dashed lines excited states.

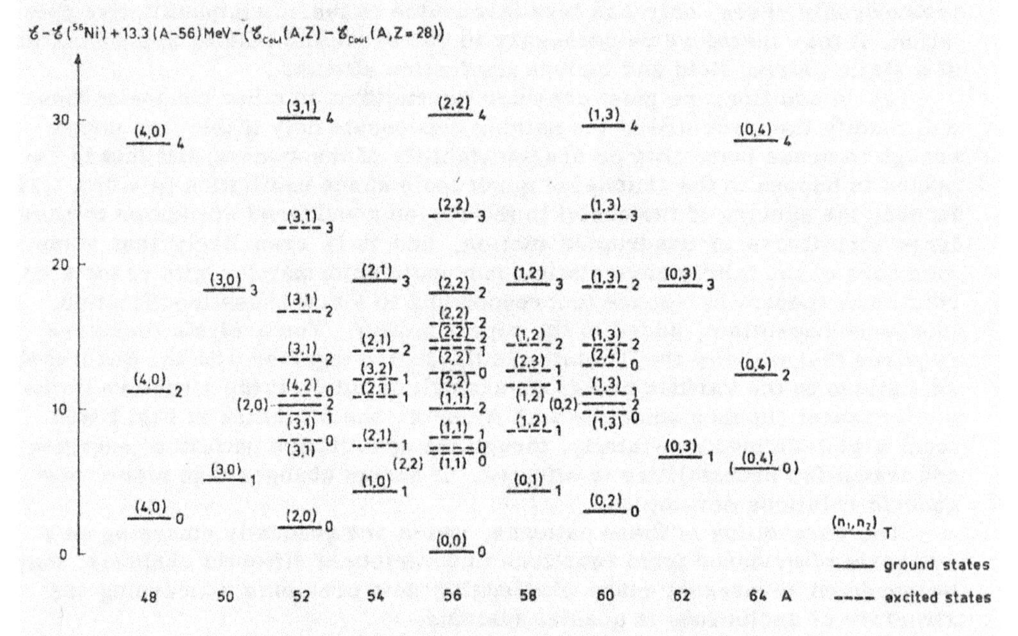

FIG.8 Schematic illustration of many-pair excitations based on ^{56}Ni.

should all be viewed together, as a single family. A number of the additional members have been found, but the evidence is still very incomplete. (For a review of the experimental data, see O. Nathan, in Nuclear Structure, Proc. Int. Symp., Dubna, (1968) 191. In trying to analyse the interaction between the quanta and developing a quantitative treatment of the spectra, at least three major effects must be considered.

(a) The first - the most conspicious in the figure - is the favouring of states with low isospin. The nucleonic interactions contain an important component with this property, which can be deduced from the symmetry term in the nuclear masses. It appears that this interaction mainly contributes to the isovector component in the nuclear monopole field (the spherically symmetric average potential), which is readily taken into account.

(b) Next, there is the pair field coupling, which implies interactions between the quanta, associated with the exclusion principle. For few quanta, these effects can be treated in terms of interactions between pairs of quanta (see Fig. 6 (f)). When many quanta are present, large anharmonicities may develop, but one can now treat the system in terms of a static pairing field, i.e. as a superfluid system. The states then organize themselves into rotational bands in isospace and in nucleon-number space, such as mentioned above. Owing to the simplicity of the pairing field, the vibration-rotation transition can be followed much more easily than for the quadrupole modes mentioned earlier and forms an instructive example of such a transition.

As regards the accuracy of such calculations, there are several points to consider. Schrieffer yesterday quoted a very comforting number for the superconductor, but in the nucleus the number of particles participating in the correlation effect is not very large. The number of doubly degenerate levels over which the diffuseness of the Fermi surface extends is about 10 in favourable cases, only 3 in less favourable cases. In a quantitative calculation, it may therefore be necessary to go beyond the leading approximation of a static pairing field and include fluctuation effects.

(c) In addition, we must consider interactions in other channels; these will modify the structure of the pairing condensate only if they are strong enough to cause instability or near-instability of the system, but this is expected to happen in the channel of quadrupole shape oscillation ($\alpha = 0$, $\lambda = 2$). Indeed, the spectra of the nuclei in the region considered are known to have large amplitudes of quadrupole motion, and it is even likely that some members of the family have static quadrupole deformations with associated rotational spectra in I-space (corresponding to a new phase modification, like ferromagentism, added to the superfluidity). The analysis therefore requires that we view the 0^+ states in the figure together with the quadrupole excitations in the various nuclei as a single family. If the structure in the $\alpha = 0$ channel changes smoothly with A and T, the 0^+ states in Fig. 8 will form a well-defined sub-family, though the quantitative pattern of energies and transition probabilities is affected. If abrupt changes take place, new generic relations develop.

The unravelling of these patterns, which are gradually emerging as a synthesis of evidence from reactions in a variety of different channels, may be expected to present many challenging new problems concerning the structure of excitations in quantal systems.

Reprint from

"NUCLEAR STRUCTURE"

INTERNATIONAL ATOMIC ENERGY AGENCY
VIENNA, 1968

PAIR CORRELATIONS AND DOUBLE TRANSFER REACTIONS

A. BOHR
THE NIELS BOHR INSTITUTE,
UNIVERSITY OF COPENHAGEN,
COPENHAGEN, DENMARK

Abstract — Аннотация

PAIR CORRELATIONS AND DOUBLE TRANSFER REACTIONS. A discussion is given of the collective modes of excitation of the nuclear pairing field, on which experimental evidence from the two-nucleon transfer reactions is accumulating. In regions of closed shells, the excitations have a vibrational character. The neutron 0^+ pair vibrations around ^{208}Pb are considered in the harmonic approximation and with the inclusion of interaction effects between the quanta of excitation. The relation between the vibrational treatment of pair correlations and the treatment in terms of superfluidity (dynamic and static pairing) is compared with the relation between vibrational and rotational descriptions of shape oscillations. The role of the isospin of the pairs is discussed with special attention to the spectrum of 0^+ pairing modes based on ^{56}Ni. When the isospin degree of freedom is included, a static pairing (superfluidity) produces a deformation in isospace, with a resulting rotational spectrum in isospace. Finally, the coupling between particles and pair vibrational quanta is briefly discussed.

ПАРНЫЕ КОРРЕЛЯЦИИ И РЕАКЦИИ ПЕРЕДАЧИ ДВУХ ЧАСТИЦ. Обсуждаются коллективные возбуждения ядерного поля спаривания, экспериментальные данные о которых собраны на основе реакций передачи двух нуклонов. Для ядер в областях замкнутых оболочек возбуждения носят колебательный характер. Рассматриваются вибрации пары нейтронов 0^+ около ^{208}Pb в гармоническом приближении и с учетом эффектов взаимодействия между квантами возбуждения. Соотношение между рассмотрением парных корреляций с точки зрения вибрации и рассмотрением с точки зрения сверхтекучести (динамическое и статическое спаривание) сравнивается с соотношением между вибрационным и вращательным описаниями колебаний формы. Обсуждается роль изоспина пар, причем особое внимание уделяется спектру 0^+ с уровнями парной природы, основанными на ^{56}Ni. Когда учитывается изоспиновая степень свободы, статическое спаривание (сверхтекучесть) дает деформацию в изопространстве, в котором появляется вращательный спектр. Наконец, обсуждается взаимодействие частиц, а также парные вибрационные кванты.

In recent years, the study of double transfer reactions has developed into an important tool for nuclear spectroscopy. It enables us to explore new aspects of nuclear structure and in particular to study the collective modes associated with nuclear pair correlations. The following paper by Nathan [1] will give a survey of the experimental findings in this new field; the present report deals with some general features of the collective pairing modes. (Many of the basic properties of pair vibrations were considered by Bès and Broglia [2]).[1]

NEUTRON PAIRING VIBRATIONS IN THE REGION OF ^{208}Pb

We first consider the neutron pairing modes based upon the closed-shell configuration of ^{208}Pb. A two-neutron transfer reaction, by which

[1] The subjects considered in the present report will be discussed in more detail in a forthcoming treatise on nuclear structure by Ben R. Mottelson and the author.

two neutrons are added to ^{208}Pb in a $J^\pi = 0^+$ state, produces the ground state of ^{210}Pb with a cross-section which is strongly enhanced as a consequence of the strong spatial correlation of the two neutrons in the ground state of ^{210}Pb. A similar enhancement characterizes the (p, t) reaction leading to the ground state of ^{206}Pb.

We may therefore view the ground states of ^{210}Pb and ^{206}Pb as collective excitations of ^{208}Pb. The quanta of excitation of these "pair vibration" modes will be labelled by the quantum number, α, the nucleon transfer number. Thus, $\alpha = +2$ for the pair addition mode and $\alpha = -2$ for the pair removal mode. In first approximation, the pair vibrations create two particles in the major shell above N = 126 or remove two particles from the shell below, but the properties of the quanta are significantly affected by the ground-state correlations in ^{208}Pb, representing the virtual excitation of pairs of particles across the gap between the shells. In particular, this zero-point vibrational motion implies an enhancement of the two-particle transfer matrix elements. The effect is of a similar nature to the enhancement of the E2 matrix elements for the low-energy quadrupole mode with $\alpha = 0$, resulting from the virtual excitation of quasiparticle pairs in the nuclear ground state.

With the $\alpha = +2$ and $\alpha = -2$ quanta as building blocks, we can construct the pair vibrational spectrum. In Fig.1, the binding energies of the states are plotted relative to that of ^{208}Pb; for convenience, a linear term in N has been added, so as to give equal energies, $\hbar\omega_2 = \hbar\omega_{-2}$, for the $\alpha = +2$ and $\alpha = -2$ quanta. Adding two or more quanta of the same kind, we obtain states denoted by $(n_1, 0)$ and $(0, n_2)$. These states correspond to the ground states of the more distant Pb isotopes. The members of the vibrational spectrum with quantum numbers $(n_1, 1)$ and $(1, n_2)$ are expected to occur at excitations of the order of 5 MeV. So far, only a single member of this type, the (1,1) level in ^{208}Pb, has been established [1, 3].

In the harmonic approximation, the energies of the vibrational excitations would equal $(n_1 + n_2)\hbar\omega$, as represented by the dotted lines in Fig.1. The observed deviations represent anharmonicity effects, which one may attempt to describe in terms of interactions between pairs of quanta. It is seen that two like quanta repel each other, while two unlike quanta exhibit a weak attraction. By determining the interaction from the states with

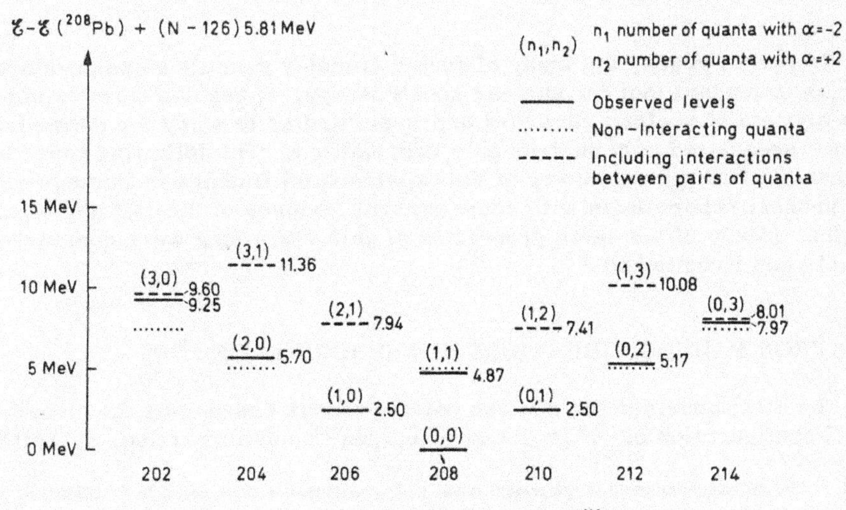

FIG.1. Neutron pairing vibrations based on ^{208}Pb.

two quanta, one can estimate the energies of the states with more quanta (dashed lines in Fig.1), and for the observed levels, the agreement is fairly good.

The interactions between the quanta receive contributions from several components of the nuclear forces. Thus, a pairing force acting among the neutrons implies a repulsion of about 400 keV between like quanta and of about 200 keV between unlike quanta [4]. The nuclear symmetry potential, which is known to contribute to the N^2 dependence of the binding energy, gives a repulsion between like quanta of about 300 keV, and a similar attraction between unlike quanta. The observed energies show that additional interactions are present; in particular, the observed small interaction between $\alpha = +2$ quanta indicates that the above-mentioned repulsive effects are partially compensated by an attractive interaction. Such an attraction may arise from the quadrupole interactions, by which the quanta can make transitions from the 0^+ state to the 2^+ state, represented by the first excited state in ^{210}Pb, or ^{206}Pb. In Fig.2, the quanta are labelled by the J-value, and one sees that an additional 0^+ level in ^{208}Pb can be formed by coupling two quanta with $J = 2$ to a resultant angular momentum, $I = 0$. The two 0^+ states repel each other through the familiar quadrupole interaction in the $\alpha = 0$ channel. The matrix element is estimated to be about 0.5 MeV, which would give a repulsion of the order of 100 keV, but the estimate is somewhat uncertain owing to lack of information on the quadrupole transition probability for the $\alpha = 2$ quantum, and of the renormalized coupling constant for the quadrupole interaction. A similar effect occurs for the states (0, 2) and (2, 0) with two identical quanta, and the rather large attractive component in the interaction of two $\alpha = +2$ quanta might indicate that the quadrupole matrix element for these quanta is especially large; information on the B(E2) value for the excitation of ^{210}Pb and on the polarization charge for a neutron in ^{209}Pb would be of significance for

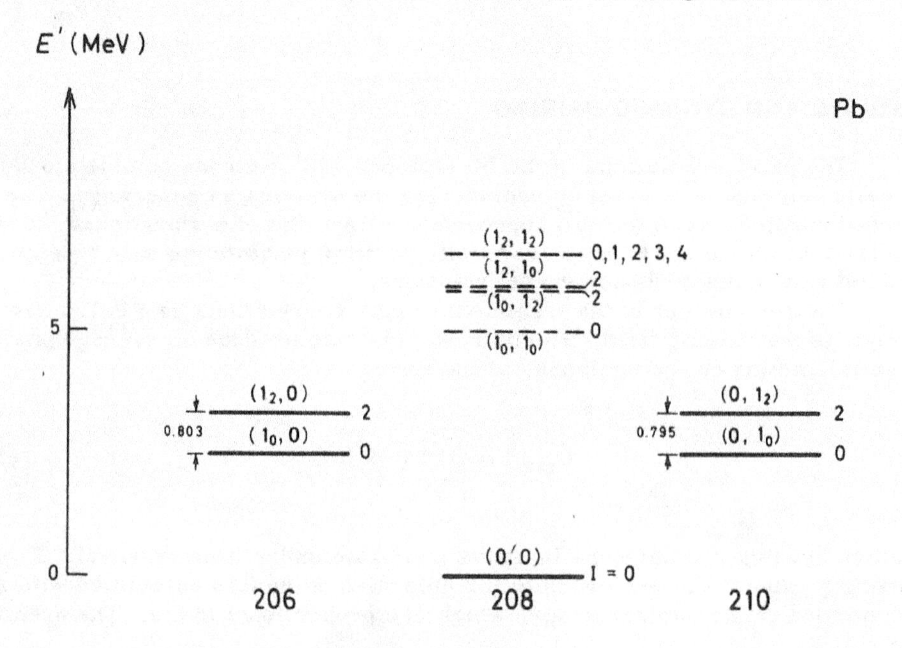

FIG.2. States in ^{208}Pb obtained by superposing quanta with $J = 0$ (1_0) and $J = 2$ (1_2).

testing this point. [There is a further structure in the 0^+ spectrum of the $\alpha = +2$ mode which may be of relevance in the present context. The calculations of the (t, p) transfer by Glendenning [5] and by Broglia and Riedel [6] predict the occurrence of a 0^+ state at about 3 MeV of excitation in ^{210}Pb with a cross-section even larger than for the ground state, and the experimental data do seem to show a very large transfer strength in this region of excitation. The occurrence of such a strong transition is connected with a subshell structure in the single particle spectrum together with the non-locality in the transfer process which gives especially strong weight to the nucleon orbits with small orbital angular momentum].

The admixing of the 0^+ states in Fig.2 also implies that the higher level can be excited by the reactions ^{206}Pb(t, p) and ^{210}Pb(p, t), and the strength of these transitions would give valuable information on the coupling.

Fig.2 also shows the two $I^\pi = 2^+$ states, which can be formed by exciting either of the quanta; the unperturbed positions of these states differ by only 8 keV and the states are therefore expected to be strongly mixed by the quadrupole interaction.

The essential property which joins the states in Fig. 1 into a collective family is the strong matrix element for two-particle transfer. In the harmonic approximation, the transition strength for the reaction $(n_1, n_2) \rightarrow (n_1+1, n_2)$ is proportional to (n_1+1). The interactions between the quanta, associated with the exclusion principle for the particles, are expected to somewhat reduce the transition strength; to first order, the correction factor has the form $(1 - \alpha n_1)$, with α a constant. The interactions also lead to a violation of the selection rule $\Delta n_1 = \pm 1$, $\Delta n_2 = 0$ or $\Delta n_1 = 0$, $\Delta n_2 = \pm 1$, and imply the occurrence of weaker transitions such as $(0, 0) \rightarrow (1, 2)$. Anharmonicity effects in the two-nucleon transfer matrix elements, for neutrons in the region of ^{208}Pb interacting by a pairing force, have been evaluated by Sørensen [4].

STATIC AND DYNAMIC PAIRING

The pair correlations in the Pb isotopes with neutrons outside closed shells can also be treated by considering the systems as superfluid. The relationship between the two approaches is like that of a vibrational and rotational treatment of shape oscillations; time permits me only briefly to indicate some of these general relations.

The key concept in the treatment of pair correlations as a collective mode is the pairing field. The pair correlations produce an average potential, acting on the nucleons, of the form

$$U_{pair} = \Delta \sum_{\nu > 0} a^+(\bar{\nu}) a^+(\nu) \tag{1}$$

which creates two nucleons in orbits conjugate under time reversal. The pairing potential is analogous to the deformed potentials associated with a distortion of the nuclear shape, which are proportional to $a^+ a$. The quantity

Δ is the deformation parameter and represents the average value of the pairing density,

$$\Delta = G\langle \sum_{\nu>0} a(\nu)a(\bar{\nu}) \rangle \tag{2}$$

which produces the potential (1).

The pairing potential violates particle number conservation just as the deformed potential violates angular momentum conservation, and the collective quantum number for the pairing modes is the particle number.

The pairing field operator is non-hermitian and Δ is therefore a complex quantity. The phase of Δ is the gauge angle, ϕ, which is conjugate to the particle number operator, and which corresponds to the orientation of the deformed nucleus.

In the ground state of ^{208}Pb, Δ oscillates about the value 0, and the excitations have a vibrational character. However, when many quanta are present, $|\Delta|$ has a mean value exceeding the fluctuations about this mean value, just as a two-dimensional oscillator in a state of large angular momentum has an amplitude whose magnitude exceeds the zero-point amplitude.

When the mean value of Δ exceeds the fluctuations, one can treat the system in terms of a static pairing field. The motion now separates into intrinsic motion, described in terms of quasiparticles (which correspond to the Nilsson particles in the deformed field), and collective motion of rotational type, the pair addition mode. The time derivative $\dot{\phi}$, which corresponds to the rotational frequency, is the chemical potential, λ, and the coriolis force is the familiar term $-\lambda N$ in the quasiparticle Hamiltonian. The gap equations for the superfluid system are the cranking formulae for the rotations in gauge space.

In the description based on a static pairing field, one neglects fluctuation effects in Δ; however, the treatment takes into account the anharmonicity in the vibrational motion, which becomes of major significance when the number of quanta is so large that the shift in λ is comparable with Δ. The pairing modes, with their especially simple structure, provide a favourable opportunity for studying the general problems associated with anharmonicity effects and transitions from vibrational to rotational spectra.

PAIRING MODES FOR Sn ISOTOPES

As an example of the structure of pairing modes in regions away from closed shells, Fig.3 illustrates the neutron pairing excitations for the Sn isotopes. The isotope ^{114}Sn marks the filling of the neutron subshell at $N = 64$, but the pairing is able to overcome this minor shell gap and establish a static pairing field. The collective transitions joining the ground states are thus of "rotational" type, and it is seen that the energies approximately follow a parabolic trajectory, rather than a linear variation as in the vibrational region around ^{208}Pb.

The dotted lines in the figure referring to the target nucleus ^{118}Sn indicate 0^+ excitations involving the degrees of freedom of the neutrons in the $50 < N < 82$ shell. The excitations represent two quasiparticle states with the spurious degree of freedom removed, and the analysis of the two particle strengths is due to Broglia et al.[7]. These excitations are seen to carry a total (t, p) strength of a few percent and a (p, t) strength of about 20% of the ground-state transition, in accordance with the analysis presented by Rasmussen [8]. Thus the subshell structure gives rise to only rather weak pairing vibrations.

Stronger pairing vibrations are expected to be associated with $\alpha = 2$ transitions into the next higher shell ($N > 82$) and $\alpha = -2$ transitions from the lower shell ($N < 50$), as indicated by the dashed lines in Fig.3. The estimate of the position of these states is rather uncertain, but it is of considerable significance to establish whether major parts of the two-particle transfer strength into the more distant levels are concentrated in well-defined collective modes.

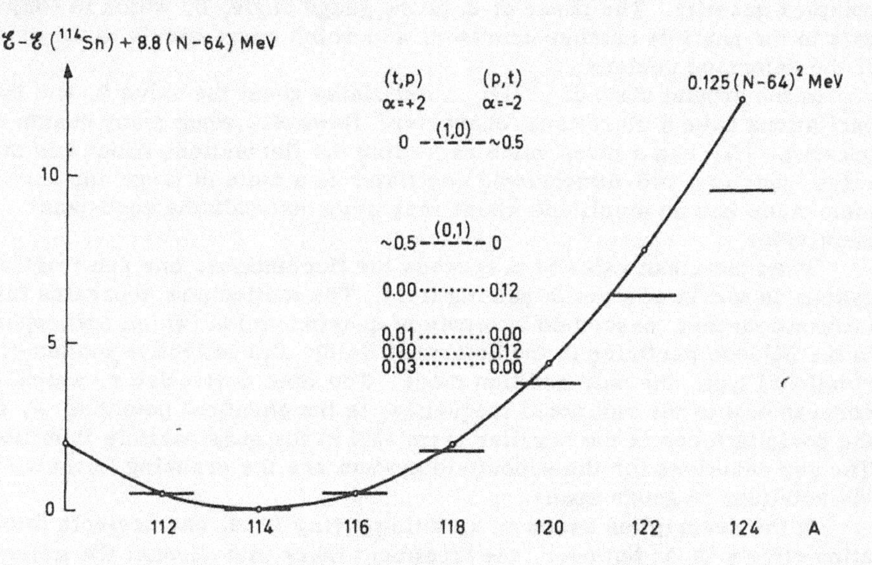

FIG.3. Neutron pair excitations in Sn isotopes. The solid lines represent ground states.

ISOSPIN OF PAIRING MODES: REGION OF A = 56

The correlated J = 0 pairs carry unit isospin and therefore involve three degrees of freedom, which gives a new dimension to the spectrum of pairing modes. We first consider the isospin structure of the pairing modes in the medium heavy nuclei with relatively small neutron excess. Taking ^{56}Ni with the closed neutron and proton shells ($N = Z = 28$) as a basis, the pair excitations again have vibrational character, and there are now six fundamental modes with $\alpha = \pm 2$, and $\mu_T = -1, 0, +1$. The one-quantum excitations have $T = 1$, and the $M_T = 1$ components (^{54}Fe and ^{58}Ni) are shown in Fig.4. Adding two quanta of the same type ($\alpha = +2$ or $\alpha = -2$), we obtain states with $T = 0$ and 2, while two unlike quanta give states of $T = 0$, 1 and 2. The $T = 2$, $M_T = 2$ component is the ground state of ^{56}Fe; the $T = 0$ and 1 levels of the configuration (1, 1) represent excited 0^+ states, which have recently been observed to be strongly populated in two-particle transfer reactions, as is discussed by Nathan [1].

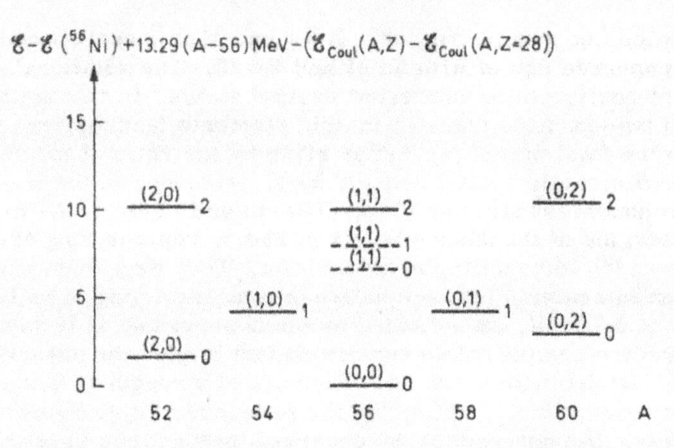

FIG.4. Pair excitations with one or two quanta added to ^{56}Ni. The quantum numbers labelling the levels are (n_1, n_2) and T.

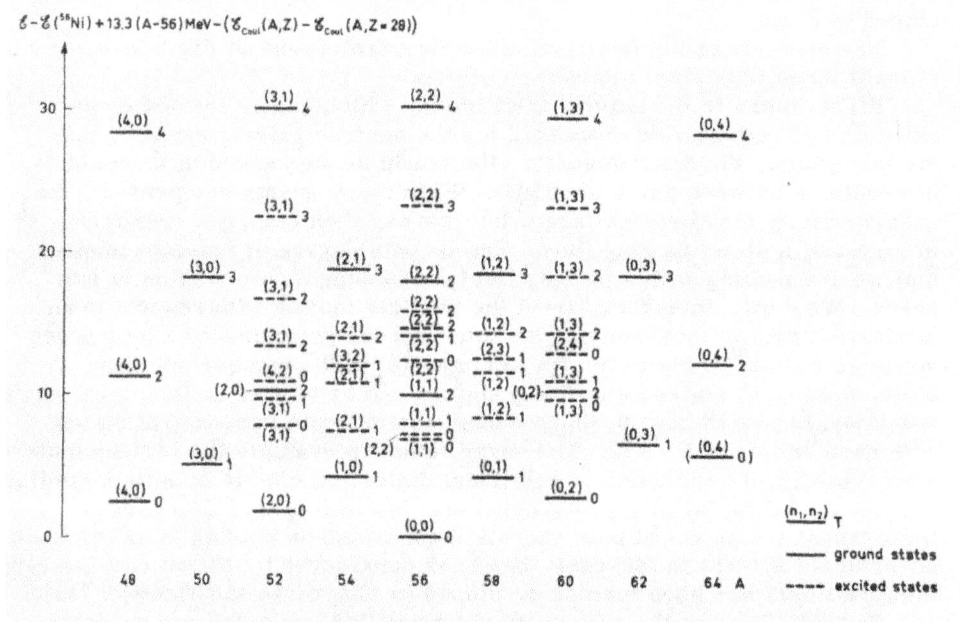

FIG.5. Schematic illustration of many-pair excitations based on ^{56}Ni.

The pattern of states obtained by superposing more quanta is illustrated schematically in Fig.5. Superposing n quanta of one type, we obtain the symmetric states T = n, n-2, ... 0 (or 1). These states have the relations of an SU(3) representation, and the various matrix elements, such as for double transfer reactions, can be expressed in terms of Clebsch-Gordan coefficients in SU(3). The $M_T = T$ components of the configurations (n_1, 0) and (0, n_2) correspond to ground states of even-even nuclei with N < 28, Z < 28, and N > 28, Z > 28, respectively.

If we add an $\alpha = 2$ quantum to a state $(n_1, 0)T_1$ with n_1 quanta of the $\alpha = -2$ type coupled to isospin T_1, we obtain states of the configuration (n_1, 1) with $T = T_1 + 1$, T_1, $T_1 - 1$. The states with maximum alignment of the isospin, i.e. with $T_1 = n_1$, $T = T_1 + 1$ and $M_T = T$ correspond to ground states of even-even nuclei with N>28 and Z< 28. The additional members of the $(n_1, 1)$ configuration represent excited states. In this coupling scheme, the ratio of two-particle transfer matrix elements leading from the target $(n_1, 0)T_1$ to the final states $(n_1, 1)T$ is given by the ratio of the vector addition coefficients $\langle T\ (M_T)_1\ \ 1\mu_\tau | T\ M_T \rangle$, where μ_τ is the transferred isospin component [+1 for (t,p), 0 for (^{3}He, p) and -1 for (^{3}He, n)].

The energies of the dashed levels in Fig.5, representing excited states, are only intended for qualitative orientation. They have been estimated by assuming an interaction between unlike quanta consisting of an isoscalar component of 0.7 MeV, as indicated by the observed (1, 1) levels (see Fig.4), and an isovector component so constructed as to give the observed positions of the $(n_1, 1)$ states with $T = n_1 + 1$. The needed strength of this component varies somewhat with n_1, reflecting the presence of higher-order interaction effects. (An analysis of the observed interactions in terms of isoscalar, isovector and isoquadrupole components has been given by Damgård [9].) Similar considerations apply to the states of the configurations $(1, n_2)$. A few of the lowest $(n_1, 2)$ and $(2, n_2)$ states are also included in Fig.5.

The analysis of the family of collective excitations in Fig.5 involves at least three important interaction effects.

First, there is the pairing interaction, which can be treated by an extension of the analysis discussed for the neutron pairing mode. Thus, for few quanta, the anharmonicity effects can be expressed in terms of interactions between pairs of quanta. When many quanta are present, the anharmonicity may become large, but one can then employ a treatment in terms of a static pairing field. The isospin degree of freedom implies that we are dealing with a pairing field possessing a deformation in isospace. We must, therefore, treat the particle motion with respect to an intrinsic frame in isospace and the total system performs rotations in isospace as well as in the gauge space conjugate to the number operator, A. Thus, the $(n_1, 0)$ states [and $(0, n_2)$ states] can be viewed as forming a single rotational band with $K_T = 0$. Such a band comprises a sequence of states with even values of $T - A/2$. Moreover, one can evaluate the various matrix elements with the inclusion of rotational distortion effects in quite a similar manner as for the ordinary rotational spectra in I-space. (A charge-independent treatment of pair correlations, based on pairing in an intrinsic co-ordinate system in isospace, has been considered by Elliott and Lea [10]; the formalism has been further developed by Ginocchio and Weneser [11].)

Second, there is the symmetry potential (the isovector component in the average nuclear potential), which gives an energy proportional to $T(T+1)$, and which is responsible for a large part of the observed splitting of the states with given (n_1, n_2).

Third, there is the quadrupole interaction. Estimates of the quadrupole interaction matrix elements of the type considered for ^{208}Pb indicate that, in the A=56 region, these matrix elements are large compared with the energies for exciting the quanta to 2^+ states; therefore, the angular momentum quantum numbers of the individual quanta are not expected to be even approximately conserved. Indeed, it is likely that among the family of excitations in Fig.5, we encounter nuclei with stable shape deformations. Evidence for deformation effects in this region of nuclei have been considered by many authors. Large deformations may especially occur for states with many quanta and low isospin. For example, it appears possible that the 0^+, 2^+, and 4^+ levels in ^{56}Ni at 4.97 MeV, 5.33 MeV, and 6.39 MeV, recently observed in the (p, t) reaction [12], may constitute a rotational band associated with the $(2, 2)T = 0$ configuration.

The occurrence of quadrupole interactions leading to instability of the spherical shape may strongly affect the pattern of the pairing modes. Thus, if abrupt changes in the nuclear shape occur with the addition of a pair of particles, the $J = 0$ double transfer strength may fractionate, as has been observed in the $N = 90$ region [13]. However, if the nuclear shape varies smoothly with A and T, the states joined together by strong $J = 0$ double transfer reactions may still form a pattern such as that illustrated in Fig.5.

For the establishment of the appropriate coupling scheme, it is of special significance to test the validity of the quantum numbers n_1 and n_2. If the deformations wash out the large gap in the single-particle spectrum at $N = Z = 28$, a static pairing field is established across this gap. In such a situation, the number of strong transitions is reduced, i.e. the pair vibrations become weak compared with the pair rotations, as illustrated by the pairing modes in the Sn isotopes (see Fig.3).

Another significant point is the degree of validity of the quantum numbers T_1 and T_2. In general, a configuration (n_1, n_2) may give rise to several states of given T, involving different values of (T_1, T_2). The extent to which these states become mixed by the interactions can be studied by testing the selection rules and intensity relations in double transfer reactions (such as discussed above), which characterize the coupling scheme in which T_1 and T_2 are constants of the motion.

The available evidence on strong $J = 0$ double transfer reactions in the region considered are reviewed by Nathan [1]. It is still very incomplete and much more data concerning transfer reactions with $J = 2$ as well as $J = 0$ will be needed to establish the appropriate coupling scheme. Also the evidence on quadrupole transitions in the $\alpha = 0$ channel (E2 transitions, inelastic scattering) must be brought into the analysis. The problem involves the familiar competition between pairing and quadrupole effects, which takes a new form for these nuclei because of the role of isospin. The development of two-particle transfer studies has given us a new powerful probe and a new way of thinking of these problems.

ISOSPIN OF PAIRING MODES: NUCLEI WITH LARGE NEUTRON EXCESS

When the isospin T is large, the coupling scheme in isospace can be treated in a "semiclassical" approximation by which the $M_T = T$ components are considered as having their intrinsic axis in isospace fully aligned in the direction of the z-axis in isospace. The pair correlation in the ground state can then be treated in terms of a pairing among neutrons and protons, separately. The modes of excitation can be characterized by the quantum number ν_τ representing the component of isospin along the intrinsic axis, and the excitations have $T = T_0 + \nu_\tau$, where T_0 is the isospin of the ground state or target nucleus.

The pairing excitations with $\nu_\tau = +1$ and -1 are the familiar ones associated with neutrons and protons, respectively, but in addition one expects pairing vibrations with $\nu_T = 0$ and $T = T_0$. The matrix elements for exciting the various M_T components are proportional to the coefficient

$$\langle T_0\ M_T = T_0\ 1\mu_T \,|\, T_0 + \nu_T,\ M_T = T_0 + \mu_T \rangle \approx \begin{cases} 1 \\ (T_0)^{-\frac{1}{2}} \\ 2^{-\frac{1}{2}}\ T_0^{-1} \end{cases} \quad \mu_T = \begin{cases} \nu_T \\ \nu_T - 1 \\ \nu_T - 2 \end{cases} \tag{3}$$

The pairing modes with $\nu_T = 0$ and $T = T_0$ are, therefore, most strongly excited in reactions with the transfer of an (np) pair, but so far there appears to be no experimental evidence concerning these modes.

INTERACTION BETWEEN PARTICLES AND PAIRING VIBRATIONS

Additional evidence with a direct bearing on the structure of the pairing modes can be obtained from the study of the interaction between single particles and the quanta of the pairing vibrations. The fundamental interaction can be illustrated by the diagrams in Fig.6a. In the Pb region, the value of the matrix element m is estimated to be about 4 (a little smaller for $\alpha = -2$, a little larger for $\alpha = 2$) while G is close to 0.1 MeV.

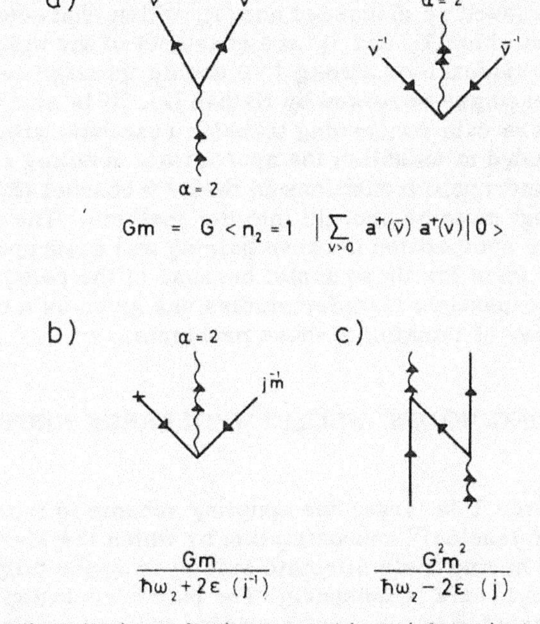

FIG.6. Diagrams illustrating the coupling between particles (or holes) and pair vibrational quanta.

First-order effects of the interaction can be studied in one-particle transfer reactions. An example is illustrated in Fig.6b; on account of the ground-state correlations in ^{208}Pb, a particle jm can be deposited in an empty orbit with the excitation of the state j^{-1} m, $n_2 = 1$ containing a hole and a quantum of the $\alpha = 2$ pairing vibration. The admixed amplitudes are of the order of 0.1 and the cross-sections thus small compared with single-particle magnitude. However, the detection of the effect would be of considerable interest as a test of the assumed coupling. The study of similar reactions in the A = 56 region would also yield very significant information on the pairing modes. In second order, the coupling gives rise to energy shifts of the states involving a particle (or hole) and a quantum (see the example in Fig.6c), but for this order it is necessary also to consider the other interaction effects discussed above, which are found to play a role in the anharmonicity of the pair vibration mode.

REFERENCES

[1] NATHAN, O., Experimental status of two-neutron transfer reactions in medium and heavy nuclei, these Proceedings.

[2] BES, D.R., BROGLIA, R.A., Nucl. Phys. 80 (1966) 289.

[3] BJERREGÅRD, J.H., HANSEN, O., NATHAN, O., Nucl. Phys. 89 (1966) 337.

[4] SØRENSEN, B., Nucl. Phys. A97 (1967), 1 and private communication.

[5] GLENDENNING, N.K. (Berkeley, California) private communication.

[6] BROGLIA, R.A., RIEDEL, C. (Niels Bohr Institute, Copenhagen) private communication.

[7] BROGLIA, R.A., RIEDEL, C., SØRENSEN, B., UDAGAWA, T., Nucl. Phys., to be published.

[8] RASMUSSEN, J., Properties of some exact solutions of the pairing force problem, these Proceedings.

[9] DAMGÅRD, J., (Niels Bohr Institute, Copenhagen) private communication.

[10] ELLIOTT, J.P., LEA, D.A., Phys. Lett. 19 (1965) 291.

[11] GINOCCHIO, J.N. (Massachusetts Institute of Technology, Cambridge, USA), WENESER, J. (Brookhaven National Laboratory, N.Y., USA) Preprint.

[12] DAVIES, W.G., KITCHING, J.E., McLATCHIE, W., MONTAGUE, D.G., RAMAVATRAM, K., CHANT, N.S. Phys. Lett. 27B (1968) 363.

[13] BJERREGÅRD, J.H., HANSEN, O., NATHAN, O., HINDS, S., Nucl. Phys. 86 (1966) 145.

Yearly photograph of institute staff and guests at The Niels Bohr Institute, on the recurrence of Niels Bohr's birthday on 7 October 1970. Courtesy of The Niels Bohr Institute, Copenhagen.

Niels Bohr Institutet

Oktober 1970

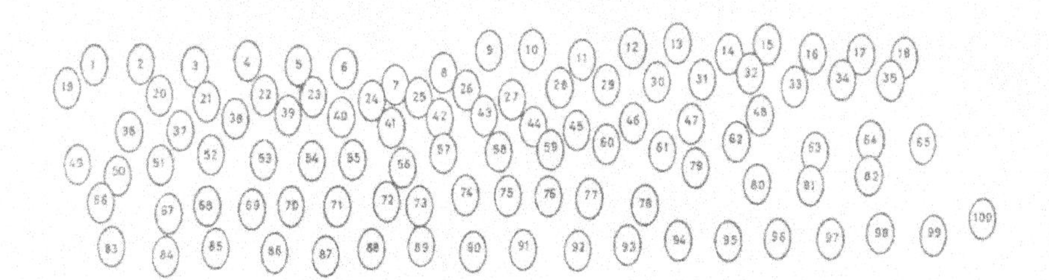

1) B. Tromborg
2) H. Nielsen
3) V.R. Pandharipande
4) B. Nilsson
5) E. Kashy
6) Z. Koba
7) Y. Munakata
8) Y. Miyachi
9) P.D. Kunz
10) E. Flynn
11) F. Michaelsen
12) F. Hansen
13) O. Holck
14) V. Møller
15) B.L. Andersen
16) F. Iachello
17) B. Olsen
18) B. Jørgensen
19) H. Johnstad
20) F. Becchetti
21) P. Hansen
22) G. Gustafsson
23) A. Holm
24) A.P. Galeao
25) S. Landowne
26) K. Jones
27) T. Kamuri
28) F. Folkmann
29) B. Rasmussen
30) P.I. Petersen
31) P. Møller Nielsen
32) V. Maarbjerg
33) F. Dickmann
34) S.G. Nilsson
35) D. Tørning
36) J. Pedersen
37) B. Herskind
38) L.E. Lundberg
39) H.J. Braathen
40) O. Johns
41) A. Sevgen
42) W. Langer
43) J. Syrak Larsen
44) B. Scharff
45) M.-L. Andersen
46) N. Brene
47) E. Perl
48) V. Paar
49) S. Barshay
50) Eva Hansen
51) B. Madsen
52) A. Lande
53) V. Alexyev
54) N.I. Manko
55) R. Broglia
56) R.Almegaard
57) K. Hansen
58) B. Sørensen
59) H. Bøggild
60) E. Dahl-Jensen
61) P. Siemens
62) J. Bondorf
63) L. Vistisen
64) E. Lohse
65) M. Willumsen
66) Sven Holm
67) B. Lenschau
68) L. Madsen
69) B. Matthiasen
70) U. Hansen
71) I. Wadil
72) B. Myglegaard
73) G. Møller Nielsen
74) H. Kiilerich
75) I. Kjems
76) B. Thorlund
77) I. Juelner
78) K. Beyer
79) I.Lundgård Rasmussen
80) N.O. Lassen
81) H. Jensen
82) N.O. Roy Poulsen
83) Jytte Sørensen
84) Lis Krogh Pedersen
85) T. Huus
86) Lise Madsen
87) M. Gavrillas
88) B. Mottelson
89) O. Nathan
90) S. Hellmann
91) Å. Bohr
92) B. Schultz
93) J. Weber
94) A. Schild
95) G. Bøggild
96) J.K. Bøggild
97) T. Gustafsson
98) B. Strømgren
99) S. Rozental
100) M. Wolf Hansen

Rotational Motion

A. BOHR

The Niels Bohr Institute, University of Copenhagen, Denmark

THE QUEST for symmetry has always been a driving force in the attempts to understand the relationships among natural phenomena. In the development of quantal physics, the scope of symmetry concepts has been greatly extended, but also the significance of symmetry breaking has come more into focus. In the exploration of nuclear phenomena, these themes have played a prominent role, and they provide a special flavour to the study of collective motion in the nucleus.

OCCURRENCE OF ROTATIONAL SPECTRA IN QUANTAL SYSTEMS

Spectra associated with quantized rotational motion were recognized in the absorption of infrared light by molecules at a very early state of development of quantum theory. The existence of rotational motion as a well-defined degree of freedom in molecules follows immediately from the existence of a semi-rigid structure represented by the atomic nuclei in their equilibrium positions.

The possible occurrence of rotational motion in nuclei became an issue in connection with the early attempts to interpret the evidence on nuclear excitation spectra (see, for example, Teller and Wheeler, 1938). The available data, as obtained, for example, from the fine structure of α-decay, appeared to provide evidence against the occurrence of lowlying rotational excitations, but the discussion was hampered by the expectation that rotational motion would either be a property of all nuclei or generally excluded, as in atoms, and by the assumption that the moment of inertia would have the classical value as for rigid rotation.

The establishment of the nuclear shell structure, which implies that the nucleons move approximately independently in the average nuclear potential, and the recognition, which came almost simultaneously, that many nuclei have equilibrium shapes deviating from spherical symmetry, created a new basis for the discussion of rotational motion in the nucleus. It was evident that the existence of a non-spherical shape implied collective rotational degrees of freedom, but one was faced with the need for a generalized treatment of rotations applicable to quantal systems that do not have a rigid structure. Such a treatment has been gradually developed over the years and the subject has continued to exhibit new facets*.

The quantal theory of rotations may find application not only to molecules and nuclei, but also to the spectra of the hadrons, as has been much discussed in recent years, in connection with the intimately related concept of Regge trajectories. Moreover, excitations of rotational character occur in the non-classical spaces that play a prominent role in quantal physics (isospace, gauge space, etc.).

The existence of a deformation, taken in the general sense of an element of anisotropy in the structure of the system, may be recognized as the hallmark of quantal systems that exhibit rotational spectra. Indeed, such an element of anisotropy is required to make it possible to specify the orientation of the system. This definition of a deformation includes the lattice-like structure in molecules and rotating pieces of solid matter; a classical physical object that may seem perfectly spherical is in fact highly anisotropic on account of the atomic constitution of matter.

The occurrence of rotational degrees of freedom may thus be said to originate in a breaking of rotational invariance. In a similar manner, the translational degrees of freedom are based upon the existence of a localized structure. However, while the different states of translational motion of a given object are related by Lorentz invariance, there is no similar invariance applying to co-ordinate frames rotating with respect to the matter distribution of the universe. The Coriolis and centrifugal forces that act in such co-ordinate frames perturb the structure of a rotating object.

In a quantal system, already the frequency of the lowest rotational excitations may be so large that the Coriolis and centrifugal forces affect the structure in a major way. The condition that these perturbations be small (adiabatic condition) is intimately connected with the condition that the zero-point fluctuations in the deformation parameters be small compared with the equilibrium value of these parameters, and the adiabatic condition

* A more detailed presentation of the considerations in the present report, as well as references to the various steps in the development of the subject, will be contained in Vol. II of *Nuclear Structure* by A. Bohr and B. R. Mottelson, Addison-Wesley/W. A. Benjamin, Inc. Reading, Mass.

provides an alternative way of expressing the criterion for the occurrence of rotational spectra.

For sufficiently large values of the angular momentum, the rotational perturbations strongly affect the structure of the system; however, rotational sequences may still occur if the properties of the system vary smoothly with the angular momentum.

ROTATIONAL DEGREES OF FREEDOM

The occurrence of a rotational spectrum corresponds to the possibility of obtaining an approximate separation of the motion represented by a total wave function of the product form

$$\Psi = \varphi_{\text{int}}(q)\, \Phi_{\text{rot}}(\omega) \tag{1}$$

where the angular variables ω specify the orientation of the system, while the co-ordinates q characterize the intrinsic motion with respect to the body-fixed frame with orientation ω.

The specification of an orientation in three-dimensional space involves three co-ordinates, such as the Euler angles, $\omega = \theta\phi\psi$, and there are three associated quantum numbers. The overall rotational invariance for the system as a whole implies the constancy of the total angular momentum I and its component M on a fixed axis. As a third angular momentum variable, one may choose the component $I_3 = K$ of the angular momentum on one of the intrinsic axes (which are labelled $\varkappa = 1, 2, 3$), but this quantity is not, in general, a constant of the motion. The full rotational degrees of freedom imply the occurrence of $(2I + 1)^2$ states for each I and $(2I + 1)$ states for each IM; the state IKM is represented by the rotation matrix $\mathscr{D}^I_{MK}(\omega)$.

The full rotational degrees of freedom come into play only if the deformation completely breaks the rotational symmetry so that it permits a unique specification of the orientation. However, the deformation may be invariant with respect to a subgroup of rotations of the co-ordinate frame. Thus, a deformation in three-dimensional space may retain axial symmetry; the deformation may also be invariant with respect to a group of finite rotations. For example, an ellipsoidal deformation is invariant with respect to rotations $\mathscr{R}_\varkappa(\pi)$ of π about any of the principal axes ($\varkappa = 1, 2, 3$).

The rotations that leave the deformation invariant constitute a subgroup G of the total group of rotations in the space considered. The well-established nuclear deformations have axial symmetry and moreover are invariant with respect to a rotation π about an axis perpendicular to the symmetry axis ($\mathscr{R}_2(\pi)$), corresponding to the invariance group D_∞. The same invariance applies to a diatomic molecule with identical nuclei.

The elements of G are part of the intrinsic degrees of freedom, and the intrinsic states can thus be classified in terms of the representations of G. For example, if the intrinsic structure involves independent-particle motion in a deformed potential, each orbit can be classified in terms of the symmetry G; in molecules, the operations contained in G can be expressed in terms of permutations of identical nuclei, in addition to the effect on the electronic orbits.

Since the elements of G are part of the intrinsic degrees of freedom, the rotational degrees of freedom are reduced correspondingly. One can express this constraint by requiring that, for each element $\mathscr{R}$ of G, we have

$$\mathscr{R}_e = \mathscr{R}_i \qquad (\mathscr{R} \in G) \tag{2}$$

where $\mathscr{R}_i$ expresses the rotation as an operator acting on the intrinsic variables, while $\mathscr{R}_e$ accomplishes the same rotation by acting on the collective orientation angles. For example, for a deformation with axial symmetry, the rotational quantum number K is constrained to have the same value as the angular momentum component of the intrinsic motion, corresponding to the absence of collective rotations about a symmetry axis. (This constraint also implies that the total wave function (1) becomes independent of the redundant third Euler angle, ψ, that specifies the orientation about the symmetry axis).

If the intrinsic state belongs to a one-dimensional representation of G, it is an eigenstate of the operations $\mathscr{R}_i$, and the rotational spectrum (1) only contains states with the same values of $\mathscr{R}_e$. If the intrinsic state belongs to a multidimensional representation of G, the total wave function is a linear combination of products of the form (1).

Thus, for a deformation of symmetry D_∞, the representations with $K = 0$ are one-dimensional and the constraint (2) for $\mathscr{R} = \mathscr{R}_2(\pi)$ implies that the rotational spectrum consists of a sequence of states with

$$(-1)^I = r \tag{3}$$

where r is the eigenvalue of $\mathscr{R}_i$ for the intrinsic state. The representations of D_∞ with $K \neq 0$ are two-dimensional ($I_3 = \pm K$, with K taken to be positive), and the wave functions take the form

$$\Psi = \left(\frac{2I+1}{16\pi^2}\right)^{1/2} \{\varphi_K(q)\, \mathscr{D}^I_{MK}(\omega) + (-1)^{I+K}\, \varphi_{\bar{K}}(q)\, \mathscr{D}^I_{M,-K}(\omega)\} \tag{4}$$

where $\varphi_{\bar{K}}$ is obtained from φ_K by the rotation $\mathscr{R}_2^{-1}(\pi)$. The band contains a state for each $I(\geqq K)$, but the phase factor in Eq. (4)

$$\sigma = (-1)^{I+K} \tag{5}$$

provides a signature dividing the band into two sequences with $\sigma = \pm 1$, each having consecutive states with $\Delta I = 2$.

The above description can be readily extended to deformations of more general type that may involve violations of other symmetries. Thus, in molecules, the deformations most frequently are unsymmetric with respect to spatial inversion; such $\mathscr{P}$ violation also occurs in nuclei, if the deformation contains components of odd multipole order. Deformations with $\mathscr{T}$ violation may occur in molecules with partial spin alignment of the electrons and also characterize the pion field produced by the nucleon spin.

A $\mathscr{P}$- or $\mathscr{T}$-violating deformation implies a two-valued collective degree of freedom; $\mathscr{P}$ violation gives rise to parity doubling and $\mathscr{T}$ violation to a doubling of states with the same values of $I\pi$. However, there may be a connection between the rotation and reflection symmetries, in the sense that the deformation may be invariant with respect to a combination of symmetries though violating the individual symmetries. For example, a molecule such as NH_3 or a nucleus with an axially symmetric octupole deformation is invariant with respect to reflection in a plane, which can be expressed as the product $\mathscr{S} = \mathscr{P}\mathscr{R}_2(\pi)$, but neither with respect to $\mathscr{P}$ nor $\mathscr{R}_2(\pi)$, separately. In such a situation, the rotational degrees of freedom are linked to that of parity so that the band contains members of both parity values ($\pi = \pm 1$) and with rotational quantum numbers satisfying the relation

$$\mathscr{S}_e = \mathscr{P} \exp\{i\pi I_2\} = \mathscr{S}_i \tag{6}$$

For example, a band with $K = 0$ contains the states

$$I = 0, 1, 2, \ldots$$

$$\pi = s(-1)^I$$

where s is the eigenvalue of $\mathscr{S}_i$ for the intrinsic state. These results are well known, but the emphasis is on a general formulation directly based on the relationship between the rotational degrees of freedom and the symmetry breaking of the deformation.

In quantal physics, attributes such as particle number and isospin that are assigned a passive role in classical physics, become dynamical operators with properties similar to those of angular momentum. In the associated angular spaces, motion of rotational type occurs when the system possesses a deformation in these dimensions.

Thus, the superfluidity in quantal many-body systems may be described in terms of deformations in the gauge space associated with particle number. These deformations are expressed in terms of densities and fields that create particles (single bosons in the case of superfluid liquid He, pairs of fermions in the case of superconductors or pair-correlated nuclei). The presence of such deformations implies the occurrence of rotation-like excitations, in which particle number takes the place of angular momentum, and for which the rotational frequency corresponds to the chemical potential.

For superfluid deformations involving a single type of particles, the orientation in gauge space is specified by a single angular variable, and the collective mode corresponds to rotations about a fixed axis. In the fermion systems, the deformation is invariant with respect to rotations of π about the gauge axis, and the condition (2) expresses the fact that the rotational excitations involve the addition or subtraction of pairs of particles.

Rotational modes of this type can be studied in superfluid nuclei and involve sequences of corresponding states, such as the ground states of even-even nuclei. A tool for directly exciting and probing the quanta of these "pair rotations" is provided by the two-nucleon transfer reaction. The corresponding mode in superconductors is involved in the Josephson junction, which can be viewed as two coupled rotors forced to rotate with respect to each other with a frequency determined by the electrostatic potential across the junction.

Pair correlations in which both neutrons and protons participate give rise to deformations in isospace as well as in the gauge space associated with total nucleon number A. The orientation thus involves four angular variables conjugate to the momentum variables ATM_TK_T. The favoured deformations in the nuclear ground states have axial symmetry in isospace and are invariant with respect to a rotation of π about an axis perpendicular to the symmetry axis in isospace followed by a rotation of $\pi/2$ in gauge space. For example, the ground states of even-even nuclei belong to a band with $K_T = 0$ that contains states with total isospin T and mass number A satisfying

$$\tfrac{1}{2}A - T = 2p \qquad (7)$$

$$p \text{ integer}$$

The empirical evidence is as yet too incomplete to test the possible occurrence of the comprehensive rotational band structure of this type.

An example of a deformation whose symmetry involves a combination of isospace with ordinary space is provided by the pion field of the nucleon. The preferred structure of the p-wave field is axially symmetric in each of the spaces and in addition is invariant with respect to arbitrary rotations performed simultaneously in space and isospace. A stable deformation of this type gives rise to a rotational band containing a sequence of states with $I = T\,(= 1/2, 3/2, \ldots)$.

These examples illustrate the great variety of symmetry patterns that may be associated with rotational motion in quantal systems.

STRUCTURE OF MATRIX ELEMENTS. EXPANSION IN POWERS OF THE ANGULAR MOMENTUM

The intimate relationship between members of a rotational band manifests itself in the regularities of the energy spectra and the intensity rules that govern transitions leading to different members of a band. The underlying deformation is expressed by the occurrence of collective transitions within a band.

For sufficiently small values of the rotational quantum numbers, the analysis of matrix elements can be based on an expansion in powers of the angular momentum. The leading-order approximation is obtained by considering the intrinsic motion for a fixed orientation. The dynamical effects of the rotation (Coriolis and centrifugal forces) give rise to perturbations of the intrinsic motion, which can be treated in terms of couplings between different bands. One can also express the rotational perturbation effects in terms of a renormalization of the effective operators acting in the unperturbed basis. Such a description corresponds to a canonical transformation that restores the coupled wave functions to the unperturbed form, now expressed in terms of new co-ordinates q, ω. The various operators, when expressed in terms of the new co-ordinates, become functions of the angular momentum. Such an approach forms a convenient basis for a phenomenological treatment exploiting the symmetry of the rotational degrees of freedom.

The matrix elements take an especially simple form for systems with axial symmetry, for which the rotational wave functions are completely specified by the quantum numbers IKM. For a spherical tensor operator $\mathcal{M}(\lambda\mu)$, the matrix elements are obtained by a transformation to the intrinsic co-ordinate system

$$\mathcal{M}(\lambda\mu) = \sum_{\nu} \mathcal{M}(\lambda\nu)\, \mathcal{D}^{\lambda}_{\mu\nu}(\omega) \tag{8}$$

In the leading-order approximation, the intrinsic moments $\mathcal{M}(\lambda\nu)$ are independent of the rotational variables (if the operator is not explicitly angular momentum-dependent), and the reduced matrix element has the form

$$\langle K'I' \| \mathcal{M}(\lambda) \| KI \rangle = (2I+1)^{1/2} \langle IK\lambda K' - K | I'K' \rangle \langle K' | \mathcal{M}(\lambda, \nu = K' - K) | K \rangle \tag{9}$$

where the last factor is an intrinsic matrix element, which is independent of I and I'. For a system with D_∞ symmetry, the matrix element between two bands, both having $K \neq 0$, may involve two terms with $\nu = |K' - K|$ and $\nu = |K' + K|$, with a relative phase involving the signature factor $(-I)^{I+K}$. While the relation (9) for large quantum numbers has a simple classical interpretation, the interference term involving the signature is a specific quantal effect, which may occur when the operator can produce effects equivalent to a finite rotation of the entire system (exchange effects).

In higher order, the renormalized intrinsic operators $\mathcal{M}(\lambda\nu)$ depend on the rotational variables through the components $I_\varkappa$ of the angular momentum with respect to the intrinsic axes and can be expanded in powers of these operators, for sufficiently small rotational frequencies. As an example, we consider a scalar operator. For matrix elements that do not involve a change in K, the effective operator is diagonal with respect to I_3 and therefore a function of $I_1^2 + I_2^2$; hence, the matrix elements are given by the familiar expansion in powers of $I(I+1)$. For matrix elements that involve a change in K, the effective operator is of the form $(I_1 + iI_2)^{\Delta K}$ multiplied by a function of $I_1^2 + I_2^2$. Thus, if we consider the energy of a band with $K \neq 0$, associated with the two-dimensional representations of D_∞ having $I_3 = \pm K$, we obtain terms with $\Delta K = 2K$ as well as $\Delta K = 0$,

$$E = E_K + AI(I+1) + BI^2(I+1)^2 + \cdots$$
$$+ (-1)^{I+K}\frac{(I+K)!}{(I-K)!}(A_{2K} + B_{2K}I(I+1) + \cdots) \tag{10}$$

In a similar manner, one can derive the expansion of matrix elements of arbitrary multipole operators.

The extensive empirical evidence that has been accumulated regarding energies and matrix elements for a variety of nuclear processes shows that, in certain rather sharply defined regions in the (N, Z) plane, the spectra exhibit the pattern characteristic of deformations with D_∞ symmetry. Examples are shown in Figs. 1–3. It is found that, in favourable cases (nuclei with large stable deformations), the power series converges rather rapidly, though in some cases there are selection rules inhibiting the leading-order intrinsic matrix elements, but allowing certain higher-order terms. The spheroidal symmetry of the deformation is directly revealed by the very large $E2$ matrix elements between members of a band.

REGGE TRAJECTORIES

For systems with axial symmetry, the rotational bands can be viewed as Regge trajectories. The analytic form of the energy, as a function of the total

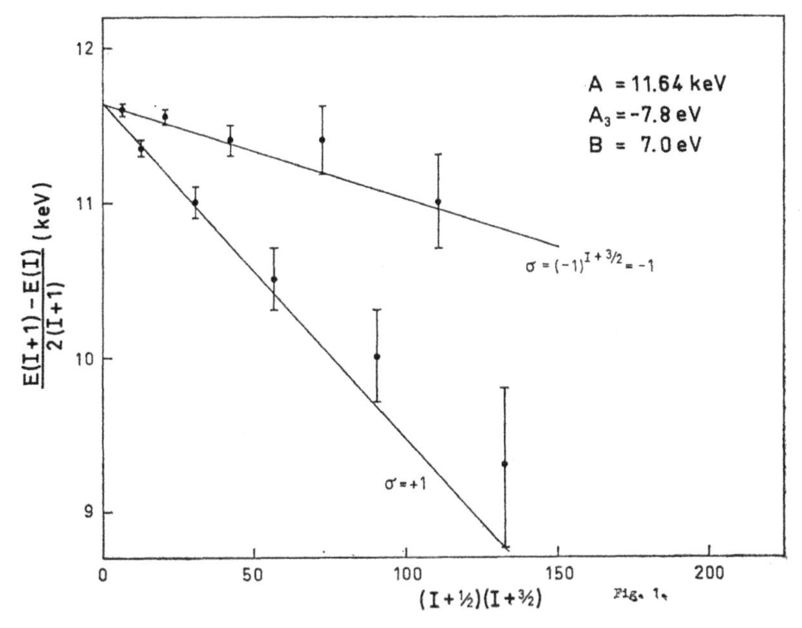

FIGURE 1 Rotational energies for ^{159}Tb. The ground-state rotational band has $K = 3/2$, and levels up to $I = 23/2$ have been populated by Coulomb excitation (Diamond *et al.*, 1963). The spectrum can be fitted by the expansion (10) with the three leading-order terms with coefficients A, B, and A_3. In the plot in the figure, such an expression corresponds to two straight lines with signature $\sigma = +1$ and -1, intersecting on the ordinate axis

angular momentum I, is given by the power series expansion (10), within the radius of convergence of this expansion. For systems with D_∞ symmetry, the presence of the signature-dependent terms in the bands with $K \neq 0$ implies that these bands involve two trajectories with $\sigma = \pm 1$.

For asymmetric systems, the bands involve a multitude of states of each I, the number increasing with I, and the problem of connecting states of different I by trajectories acquires new aspects. As an illustration, Fig. 4 shows the solution to the asymmetric rotor Hamiltonian

$$H_{\text{rot}} = \sum_{\varkappa=1}^{3} A_\varkappa I_\varkappa^2$$
$$A_\varkappa \equiv \frac{\hbar^2}{2\mathscr{J}_\varkappa} \tag{11}$$

for a particular choice of the asymmetry parameter characterizing the relative values of the inertial parameters. The Hamiltonian matrix is written in a representation in which one of the components of $I_\varkappa$ is diagonal; in this

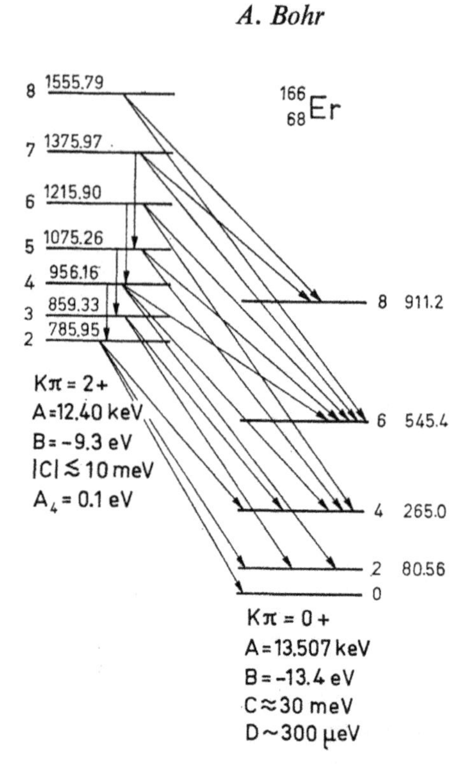

FIGURE 2 Ground-state band ($K\pi = 0+$) and excited $K\pi = 2+$ band in ^{166}Er. The energies of the ground-state band are taken from the Table of Isotopes by Lederer *et al.*, 1967, while the energies of the $K\pi = 2+$ band (γ-vibrational excitation) are from C. W. Reich and J. E. Cline, private communication. The energies of the bands can be fitted by the expansion (10) with the parameters shown in the figure. The arrows represent observed $E2$ transitions

matrix, the total angular momentum I is considered to be a parameter that can be continuously varied. For non-integer values of I, the matrix does not break off and is of infinite dimension. The trajectories in Fig. 1 show the eigenvalues of these infinite matrices as a function of I in the representation associated with the axes that have the largest and smallest moment of inertia (labelled by $\varkappa = 1$ and 3, respectively). For the intermediate axis ($\varkappa = 2$), the Hamiltonian matrix has no discrete spectrum for non-integer I. The $\varkappa = 1$ and $\varkappa = 3$ trajectories intersect at the physical values of I, but are seen to connect different sequences of states in the total two-dimensional array of levels in the band. One may attempt to use the transition probabilities to characterize dominant trajectories. For an ellipsoidal deformation, it is found that the strongest $E2$ transitions follow the $\varkappa = 3$ trajectories in the lower part of the spectrum, and the $\varkappa = 1$ trajectories for the highest states of given I.

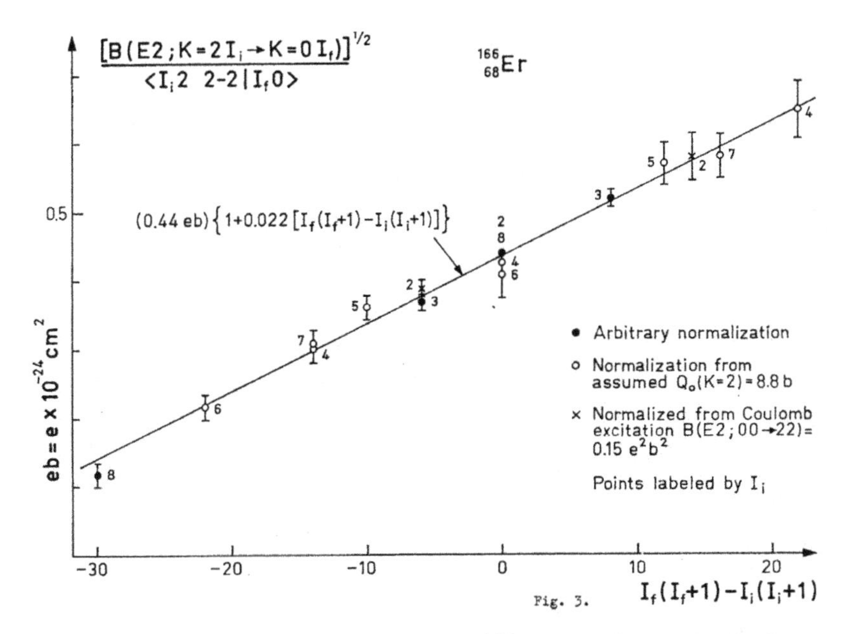

FIGURE 3 $E2$-transition amplitudes in ^{166}Er. The figure shows the measured transition probabilities for the $E2$ transitions between the $K\pi = 2+$ and $K\pi = 0+$ band shown in Fig. 2. The figure is based on the γ intensities measured by Gallagher *et al.* (1965) and by Günther and Parsignault (1967). The leading-order intensity relation (9) corresponds to a constant value for the quantity plotted in the figure, while the inclusion of the leading-order rotational coupling effects yields a straight line

The analysis of the asymmetric rotor raises questions concerning the general significance of one-dimensional Regge trajectories and the conditions under which trajectories can be defined in systems that do not allow of a separation between intrinsic and rotational motion. (For a further discussion of the analytic structure of the solutions to the asymmetric rotor Hamiltonian, see Talman, 1971).

INTERPRETATION OF PHENOMENOLOGICAL PARAMETERS

The phenomenological description of rotational spectra, as discussed in the preceding sections, provides a basis for the analysis of the experimental data in terms of the physically significant parameters that characterize the intrinsic structure of the system and its coupling to the rotational motion.

For the nuclear systems, one may attempt an interpretation of these parameters, starting from independent-particle motion in a potential of self-consistent shape. In the description of the intrinsic motion, one must include

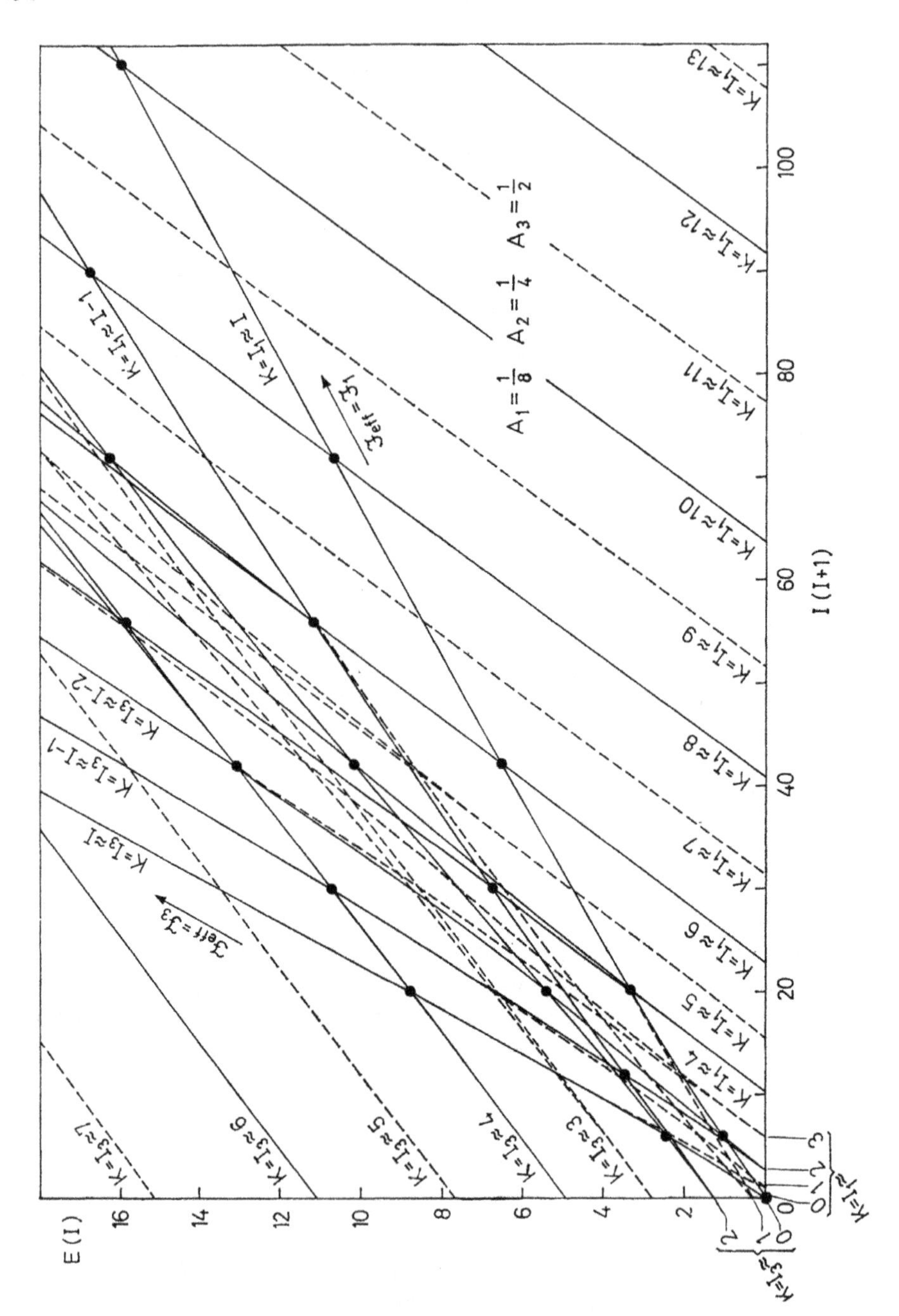

FIGURE 4 Trajectories for asymmetric rotor. The full-drawn lines are the trajectories associated with even values of K, while the trajectories with odd K are dotted. The small filled circles at the intersections of two full-drawn trajectories represent the physical states with the symmetry quantum numbers $(r_1 r_2 r_3) = (+1\ +1\ +1)$, where $r_\varkappa$ is the eigenvalue of the rotational wave function for the operation $\mathscr{R}_\varkappa(\pi) = (-1)^{I_\varkappa}$

the pair correlations as well as further interaction effects such as those that give rise to collective vibrational modes. On this basis, it is possible to account in rather great detail for the observed sequence and quantum numbers of the intrinsic excitations as well as for many features of the matrix elements associated with the I-independent terms in the effective operators.

The I-dependent terms represent the effect of the rotation on the motion of the particles, and may be analyzed in terms of the Coriolis and centrifugal forces acting in a co-ordinate frame rotating with given frequency ω_{rot}. Such an analysis gives the properties of the system as a power series in ω_{rot}, and the connection between rotational frequency and angular momentum can be obtained from the canonical relation

$$\hbar(\omega_{rot})_\varkappa = \frac{\partial H}{\partial I_\varkappa} \tag{12}$$

This approach has been rather successful, in particular in accounting for the observed nuclear moments of inertia that are typically about a factor of two smaller than the values for rigid rotation.

The simple interpretation of the spectra of the deformed nuclei and the powerful tools available for the precision study of nuclear properties provide the opportunity for a very detailed exploration of the dynamics of the rotational motion in nuclei. The evidence indicates that, at least in some cases, the estimated Coriolis coupling acting on an unpaired particle is too large by about a factor of two. This evidence comes from the observed difference in moments of inertia of odd-A and even-even nuclei as well as from the effect of the Coriolis coupling on $E2$ matrix elements between bands with $\Delta K = 1$ in odd-A nuclei (see, for example, Stephens *et al.*, 1968). The finding may seem surprising in view of the adequacy of the Coriolis coupling in accounting for the total moment of inertia of even-even nuclei, and it is important to obtain more detailed evidence on these phenomena, since we are here dealing with the basic coupling between collective rotations and the motion of individual nucleons.

The interpretation of the I-dependent properties in terms of particle motion in a rotating potential (the so-called cranking model, first employed by Inglis, 1954) may appear to involve a semi-classical element. The basis of the approach has been much debated and there have been many alternative formulations that have elucidated the quantal features of the rotational motion (see, for example, Tomonaga, 1955; Peierls and Yoccoz, 1957; Bohr and Mottelson, 1958; Thouless and Valatin, 1962; Peierls and Thouless, 1962; Kerman and Klein, 1963; Villars, 1965; Marshalek and Weneser, 1969; Beliaev and Zelevinskij, 1970). A derivation that emphasizes the quantal basis of the treatment can be obtained by starting from particle motion in a deformed static potential, $V_0(r, \vartheta)$, and adding interactions that restore the rotational symmetry. These interactions can be described in terms of the coupling between the particle motion and the collective degree of freedom associated with variations in the orientation. Thus, a small change $\delta\theta$ in orientation produces an added potential

$$\delta V = -\frac{\partial V_0}{\partial \vartheta}\delta\theta \tag{13}$$

acting on the individual particles. The generation of the collective mode itself through this interaction can be treated in the same manner as the analysis of collective vibrational modes in terms of the particle-vibration coupling (see Mottelson, 1970). In the usual normal mode (or random phase) approximation, one obtains a collective mode of zero frequency with an inertial parameter equal to that given by the cranking model. Such a treatment of the rotational motion as a vibrational mode is justified for a particle-rotation interaction slightly smaller than the value (13) required by rotational invariance, in which case the nucleus oscillates with large amplitude about the orientation of the potential V_0. The slight increase in the interaction that restores the rotational invariance changes the vibrational into a rotational mode, but is not expected to significantly affect the inertial parameter.

The additional normal modes emerging from the treatment of the coupling (13) give the intrinsic spectrum of $K\pi = 1+$ excitations. Thus, the analysis at the same time leads to a removal of the spurious degrees of freedom that is present in the description of the intrinsic motion in terms of the complete spectrum of excitations for particle motion in a static deformed potential. The problem of the $K\pi = 1+$ intrinsic excitation spectrum and the possible occurrence of collective effects in this channel is closely related to the structure of the rotational motion, and empirical evidence on these modes of excitation may elucidate the problem encountered in the analysis of Coriolis coupling effects mentioned above.

The treatment of the interaction (13) also yields expressions for the collective orientation angles, in terms of the degrees of freedom of the particle motion in a potential of fixed orientation. In general, these collective variables, considered as functions of the position, momentum, and spin variables of the nucleons, have a complicated structure, corresponding to the fact that this functional relationship reflects the detailed dynamics of the rotational motion.

A very special situation arises if all the particle excitations generated by the interaction (13) have the same frequency, as for particle motion in a harmonic oscillator potential, if we consider only excitations that shift a quantum of oscillation from one direction to another and neglect the high-frequency excitations involving the creation of two quanta. From the relation

$$\frac{\partial V_0}{\partial \vartheta} = i[I_1, H_0] \tag{14}$$

where I_1 is the angular momentum conjugate to ϑ and where H_0 is the Hamiltonian for particle motion in the fixed deformed potential, it follows that, when all the excitations generated by $\partial V_0/\partial\vartheta$ have the same frequency, the operators $\partial V_0/\partial\vartheta$ and I_1 are proportional to each other. Since I_1 is the generator of infinitesimal rotations, we may conclude that the perturbations produced by $\partial V_0/\partial\vartheta$ can be described in terms of a superposition of states representing the given initial configuration oriented in different directions. The weighting of the different orientations can be obtained by exploiting the rotational invariance of the total Hamiltonian. Thus, for an axially symmetric configuration with $I_3 = K$, the total wave function has the form

$$\Psi_{IKM} = \text{const} \int \varphi_0(K; \omega)\, \mathscr{D}^I_{MK}(\omega)\, d\omega \tag{15}$$

This projection integral is equivalent to a wave function of the form (1) and includes rotational perturbations to all orders.

The special situation considered corresponds to the U_3 model of Elliott (1958), which played such an important role in clarifying the relationship between rotational and individual-particle motion. More generally, one may consider the projected state (15) as the zero'th order approximation to the wave function expressed in terms of the co-ordinates of the individual particles. However, when the intrinsic spectrum contains several frequencies, the dynamic effects of the rotation are not adequately included in the state (15).

ROTATIONAL MOTION FOR LARGE QUANTUM NUMBERS. DISCONTINUITIES IN ROTATIONAL BANDS

In the preceding, we have focussed attention on the properties of bands for values of the rotational quantum numbers that are sufficiently small to permit a description in terms of a power series expansion in the rotational frequency about the value zero. For large rotational quantum numbers, such a description becomes inadequate, even though the properties of the band vary smoothly with the rotational quantum number. Such a situation is encountered in the pair rotations for which the rotational frequency (the chemical potential) always has a significant influence on the intrinsic motion.

The familiar treatment of the pair correlations in terms of the Hamiltonian

$$H' = H - \lambda n \tag{16}$$

where λ is the chemical potential and n the particle number, is the analogue of the cranking model. The term $-\lambda n$, which is conventionally viewed as a Lagrange multiplier term associated with the constraint of fixed average particle number, acquires a dynamical interpretation in terms of the Coriolis coupling acting in the co-ordinate frame rotating in gauge space.

In recent years, there has been considerable progress in the study of the domain of convergence of the power series expansion around $I = 0$ for the rotational spectra in heavy nuclei. In many cases, it has been possible to follow the rotational bands to values of I close to 20. As illustrated by the examples in Figs. 1 and 2, the ratio of the coefficients B/A that characterize the leading-order deviations from the $I(I + 1)$ dependence of the rotational energies is in favourable cases of the order of 10^{-3}. This might indicate a radius of convergence for the power series expansion of $I \approx 30$, but the magnitude of the higher-order terms implies a considerably poorer rate of convergence. However, it has been noted (Harris, 1965; Mariscotti *et al.*,1969) that the rate of convergence is significantly improved if the energy is expressed as a power series in the rotational frequency rather than in the angular momentum. An example is illustrated in Figs. 5a and 5b, which shows the energies of the ground-state band of ^{172}Hf. Already for $I = 8$, the energy shows significant deviations from the expansion to fourth order in I, but the corresponding frequency plot, which gives the moment of inertia as a function of the square of the rotational frequency, can be represented by a straight line over a considerably wider domain. The moment of inertia is here defined as [see Eq. (12)]

$$\mathscr{I} \equiv \frac{\hbar I}{\omega_{\text{rot}}} = \frac{\hbar^2}{2}\left(\frac{\partial E}{\partial I(I+1)}\right)^{-1} \tag{17}$$

One can also exhibit the faster convergence of the frequency expansion by noting that an energy expression containing only terms of second and fourth order in the frequency,

$$E_{\text{rot}} = \alpha\omega_{\text{rot}}^2 + \beta\omega_{\text{rot}}^4 \tag{18}$$

implies relations between the higher coefficients in the expansion in powers of $I(I + 1)$

$$\begin{aligned} \frac{C}{A} &= 4\left(\frac{B}{A}\right)^2 \\ \frac{D}{A} &= 24\left(\frac{B}{A}\right)^3 \end{aligned} \tag{19}$$

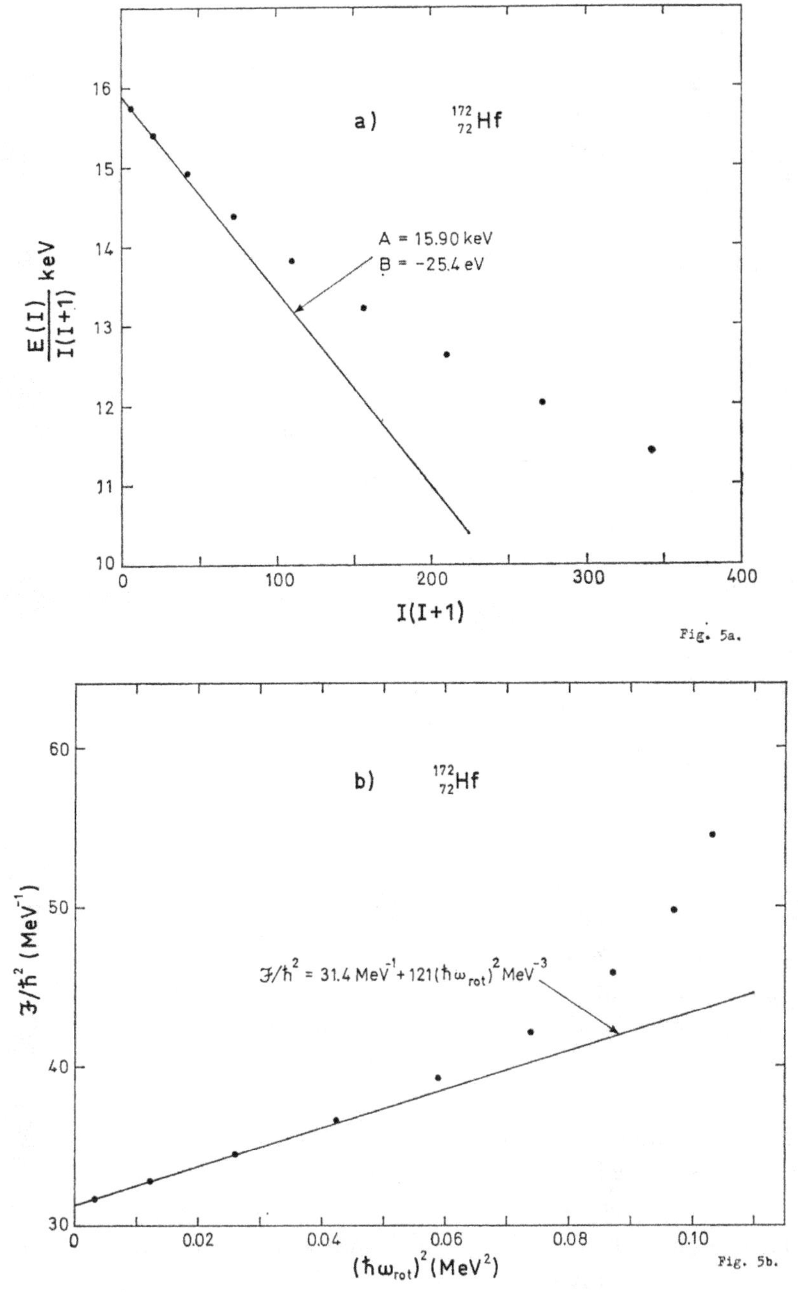

FIGURE 5 Rotational energies of the ground-state band in ^{172}Hf. The rotational levels have been studied by means of the reaction ^{165}Ho(^{11}B, 4*n*) by Stephens *et al.* (1965), and by means of the reaction ^{171}Yb(α, 3*n*) by S. Jägare, S. A. Hjorth, H. Ryde, and A. Johnson (priv. com.). In Fig. 5a, the energy is plotted as a function of $I(I+1)$, and a straight line corresponds to the expansion (10) with the leading-order terms proportional to A and B. In Fig. 5b, the moment of inertia is plotted as a function of the rotational frequency, on the basis of the relations (12) and (17)

Table 1 contains examples where precision measurements of the rotational energies have enabled the determination of the coefficients C and D; it is seen that the relations (19) are rather well fulfilled. An attempt to determine the higher-order terms in the frequency expansion (18) shows that the coefficient γ of $\omega^6{}_{\rm rot}$ is appreciably smaller (by a factor of five or more) than β^2/α. Theoretical estimates of the coefficient γ have not so far provided an explanation of this striking feature of the expansion.

TABLE 1 Test of relations implied by linear dependence of $\mathscr{I}$ on $(\omega_{\rm rot})^2$
The rotational energy expansion coefficients A, B, C, and D are calculated from the energies of the $I = 0, 2, 4, 6$, and 8 members of the ground-state rotational band. The data are taken from the compilations by Groshev *et al.* (1968) and (1969) and the more recent measurements on ^{156}Gd and ^{158}Gd (H. R. Koch, priv. com.)

Nucleus	A keV	$-\frac{B}{A} \times 10^3$	$\frac{C}{A} \times 10^6$	$4\left(\frac{B}{A}\right)^2 \times 10^6$	$-\frac{D}{A} \times 10^9$	$-24\left(\frac{B}{A}\right)^3 \times 10^9$
^{156}Gd	15.0332	2.362	14.3	22.3		
^{158}Gd	13.3353	1.069	4.10	4.56		
^{162}Dy	13.5174	0.934	3.74	3.49	16	20
^{164}Dy	12.2858	0.738	1.44	2.18		
^{168}Er	13.3431	0.547	1.39	1.20	4.2	4.0
^{178}Hf	15.6203	1.005	4.14	4.04	22	24

For the largest values of I that have been reached in the heavy deformed nuclei, the rotational energies show a behaviour that is not well described by an expansion about zero frequency and which suggests that the nucleus is undergoing some major modification as a result of the perturbations produced by the rotational motion (see, for example, Fig. 5b). It is possible that one is encountering a phase transition from a pair-correlated or superfluid state to a normal state without pair correlations, which is expected to rotate with the rigid-body value for the moment of inertia. Considerable interest attaches to the further exploration of the spectrum in this domain; the nucleus offers the opportunity for a study of individual quantum states associated with such transitions, which may complement the information obtained from the study of phase transitions in macroscopic systems.

An example of a discontinuity in rotational sequences of an especially simple structure is provided by the breaks in the pair-rotational bands associated with the shell closings. Nuclei with filled major shells have no stable superfluid deformation; otherwise the shell structure would be washed out by the pair correlations. With the approach to the closed-shell configuration, the pair-rotational bands therefore become strongly coupled to the intrinsic degrees of freedom associated with the fluctuations in the pair field and, in the region of the closed-shell configurations, the excitations of the pair field acquire a vibrational character. Thus, the bands coming from one side of the discontinuity may be continued through a pair vibrational spectrum to a merging with the bands at the other side of the discontinuity. The rotational frequencies and moments of inertia at the two sides of the shell closing are shifted by amounts depending on the energy gap between the shells.

A somewhat similar discontinuity may occur in the ordinary rotational spectra in light nuclei, in situations where the energy gap between the shells prevents the particle orbits from increasing their angular momentum in response to the rotational motion. In such a situation, the band terminates when the angular momenta of the particles have been fully aligned to the extent permitted by the exclusion principle. This type of discontinuity in the band structure was first exhibited by the U_3 model (Elliott, 1958); the alignment effect can be studied more generally by an analysis, along the lines indicated above, of the rotational perturbations caused by the Coriolis interaction. In many of the light nuclei, the bands have been followed to values of I equal to the predicted maximum value I_{max}, but the evidence for a discontinuity in the band structure is still rather inconclusive, and the nature of the relationship to the states of higher I that can be obtained by shifting particles into higher shells has not so far been explored. In heavy nuclei, the values of I_{max} are very large ($\gtrsim$ 100) and are not only far beyond the domain that has been explored, but may also exceed the values of the angular momentum at which the nucleus becomes unstable with respect to fission.

The study of nuclear states with large angular momenta is rapidly advancing due, in particular, to the advent of heavy ion beams that can transfer large amounts of angular momentum into the nucleus (50 to 100 units). In such processes, the highly excited compound nucleus rapidly cools down to the neighbourhood of the "yrast" line, representing the lowest levels of given I. In the example considered in Fig. 5 (^{165}Ho(^{11}B, 4n)^{172}Hf), compound nuclei are produced with angular momenta up to $I \approx 35$, and only rather few units of angular momentum are expended before the nucleus reaches to within about an MeV of the yrast line. The further decay follows rather close to the yrast line until the main intensity flows into the ground-state band at $I \sim$ 14–16. At slightly higher values of I, it appears that the ground-state band merges with other bands.

The nuclear spectrum in the region of the yrast line, above $I \sim 20$, remains virgin territory and involves nuclear matter in a new form. Though highly excited, the nucleus is "cold", since the energy is mainly concentrated in a single or a few collective degrees of freedom that describe the modification in the nuclear structure produced by the Coriolis and centrifugal forces.

The tentative evidence provided by the heavy ion-induced reactions appears to imply the systematic occurrence of enhanced $E2$ transitions leading along paths close to the yrast line (Diamond *et al.*, 1969), but the nature of the collective motion is not yet established. One possibility may be a stabilization of the rotational motion through the development of a triaxial shape. The rotational distortion effects tend to break the axial symmetry, thereby removing the degeneracy between the moments of inertia perpendicular to the symmetry axis. The system can exploit the asymmetry by rotating about the axis with the largest moment of inertia. (The rotational coupling effects that produce the asymmetry are the same as those responsible for the coupling between the ground-state and γ-vibrational bands implied by the I-dependent term in the $E2$ matrix elements in Fig. 3).

REFERENCES

S. T. Belyaev and V. G. Zelevinsky (1970), *Yadernaja Fizika* **11**, 741 (1970)

A. Bohr and B. R. Mottelson, *Kgl. Norske Vid. Selsk. Forhandlinger*, **31**, No. 12 (1958)

R. M. Diamond, F. S. Stephens, W. H. Kelly, and D. Ward, *Phys. Rev. Letters* **22**, 546 (1969)

J. P. Elliott, *Proc. Roy. Soc.* (*London*), **A245**, 128 and 562 (1958)

C. J. Gallagher, Jr., O. B. Nielsen, and A. W. Sunyar, *Phys. Letters* **16**, 298 (1965)

L. V. Groshev, A. M. Demidov, V. I. Pelekhov, and L. Sokolovskij, *Nuclear Data* **5**, 1 (1968) and **5**, 243 (1969)

C. Günther and D. R. Parsignault, *Phys. Rev.* **153**, 1297 (1967)

S. M. Harris, *Phys. Rev.* **138**, B509 (1965)

D. R. Inglis, *Phys. Rev.* **96**, 1059 (1954)

A. K. Kerman and A. Klein, *Phys. Rev.* **132**, 1326 (1963)

C. M. Lederer, J. M. Hollander, and I. Perlman, *Table of Isotopes*, Sixth Ed., John Wiley and Sons, New York, 1967

M. A. J. Mariscotti, G. Scharff-Goldhaber, and B. Buck, *Phys. Rev.* **178**, 1864 (1969)

E. R. Marshalek and J. Weneser, *Ann. Phys.* (*N.Y.*) **53**, 569 (1969)

B. R. Mottelson, Report Solvay Conference 1970; this volume

R. E. Peierls and J. Yoccoz, *Proc. Phys. Soc.* (London) **A70**, 381 (1957)

R. E. Peierls and D. J. Thouless, *Nuclear Phys.* **38**, 154 (1962)

F. S. Stephens, N. L. Lark, and R. M. Diamond, *Nuclear Phys.* **63**, 82 (1965)

F. S. Stephens, M. D. Holtz, R. M. Diamond, and J. O. Newton, *Nuclear Phys.* **A115**, 129 (1968)

J. D. Talman, *Nuclear Phys.* **A 161**, 481 (1971)

E. Teller and J. A. Wheeler, *Phys. Rev.* **53**, 778 (1938)

D. J. Thouless and J. G. Valatin, *Nuclear Phys.* **31**, 211 (1962)

S. Tomonaga, *Prog. Theor. Phys.* **13**, 467 (1955)

F. M. H. Villars, *Nuclear Phys.* **74**, 353 (1965)

Reprinted from
2 April 1971, Volume 172, pp. 17-21

Concepts of Nuclear Structure

Aage Bohr

2 April 1971, Volume 172, Number 3978 SCIENCE

Concepts of Nuclear Structure

Aage Bohr

The three or four decades since Charles Lauritsen (*1*) together with other pioneers opened the field of nuclear structure and nuclear reactions to systematic study have witnessed an astounding development and precision of the experimental tools for these studies. I shall not attempt to describe the tremendous richness of phenomena that have become accessible through this development, and the far-reaching ramifications in many fields of science; but I should like to comment on some of the concepts that have played a role in the interpretation of the nuclear phenomena and some of the issues that have been involved (*2*).

Quantal Theory and the Many-Body Problem

The nucleus is an example of a quantal many-body system. At all the steps of the quantum ladder, from the macromolecules and matter in bulk to the elementary particles, one deals with composite systems or more generally with the problem of how the quanta themselves are built out of other quanta. The many-body problem in its different manifestations is a common theme in the current broad development of quantal physics.

The early phases of quantal theory gave emphasis to phenomena which could be described in terms of one-particle motion or a few simple degrees of freedom or, as in electrodynamics, in terms of a perturbation in the motion of free quanta, and which could therefore be directly mastered by a solution of the basic equations of motion. In the study of the many-body systems, however, the situation has been a different one. The structural possibilities and the variety of correlation effects that may occur in a system such as the nucleus are so vast that a crucial problem has been the identification of the appropriate concepts and degrees of freedom to describe the phenomena that one encounters. Progress in this direction has been achieved by a combination of many different approaches, including the clues provided by experimental discoveries, theoretical studies of model systems, and the establishment of general relations following from considerations of symmetry.

In the developing understanding of nuclear structure, the attempts to achieve a proper balance between independent-particle and collective degrees of freedom has been a recurring and central theme. This may be true of all many-body systems, but, in the nucleus, because of the possibility of detailed studies of individual quantum states, this issue was encountered in an especially concrete form.

The earliest discussions of nuclei, as built out of neutrons and protons, were based on an independent-particle picture, similar to that which had been successfully employed in the analysis of atomic structure. However, the pioneering studies of nuclear reactions soon revealed features that could not be comprehended on this basis. Especially striking was the frequent occurrence of resonance capture of incident nucleons. The earliest evidence for such processes was obtained by Lauritsen and his co-workers in the Kellogg Laboratory.

The startling new situation had a strong appeal to my father's imagination, and he was able to bring the available experience together into a picture of the nuclear dynamics that gave emphasis to the strong coupling in the motion of the nucleons. On this basis, the course of the nuclear reactions could be understood in terms of the formation of a compound nucleus involving the excitation of a large number of degrees of freedom, and many features of its decay could be described by statistical concepts. The basic modes of excitation of the nucleus envisaged in this picture are of collective vibrational character, similar to the oscillations of a liquid drop. A few years later, a striking example of collective nuclear motion was provided by the discovery of the fission process.

On this background, the definite establishment of nuclear shell structure and its interpretation in terms of a long mean free path for motion of a nucleon in the nucleus came as another big surprise. One was then faced with the problem of reconciling the occurrence of single-particle and collective degrees of freedom and of exploring their mutual relationship.

Many phenomena could be immediately understood in terms of the interplay of the two components in the nuclear dynamics. An especially striking effect, first noted by Rainwater, is the tendency of the nucleons in unfilled shells to produce large spheroidal deformations in the nuclear shape.

While the need to include both single-particle and collective degrees of freedom in the description of the nucleus was apparent, one faced problems arising from the fact that such a description employs an overcomplete set of variables. Thus, it was recognized that the particle degrees of freedom

The author is professor of physics at the Niels Bohr Institute, University of Copenhagen, Copenhagen, Denmark. This article is adapted from his address at the first annual C. C. Lauritsen Memorial Lecture, at California Institute of Technology, Pasadena, on 29 October 1970.

must be significantly modified by the presence of the collective modes, and that in turn the shell structure could have a profound effect on the collective motion.

Nuclear Field Theory

In the development that has led to a gradual clarification of these issues, the concept of the single-particle field (or potential) generated by a collective deformation has played a central role. The field is, to a first approximation, proportional to the deformation amplitude α, and gives a contribution to the nucleonic energy of the form

$$H' = \kappa\alpha\, F(x)$$

where κ is a coupling constant while $F(x)$ gives the dependence of the potential on the nucleonic variables. The energy H' constitutes a coupling between particle and vibrational motion and is analogous to the electron-phonon and electron-plasmon interactions in a metal and to the particle-phonon coupling in liquid helium.

In terms of the particle-vibration coupling, one can develop a systematic treatment of the elementary modes of excitation in the nucleus. The first-order effects of the coupling can be illustrated by Feynman diagrams, such as

representing inelastic scattering of a particle with the excitation of a phonon and the decay of a collective phonon into a particle-hole pair. In higher orders, one obtains effective interactions between two particles, between a particle and a phonon, anharmonicity effects in the vibrational motion, and the like, corresponding to diagrams such as

At the same time, the coupling emerges as the mechanism that organizes the collective motion of the particles, as illustrated by diagrams such as

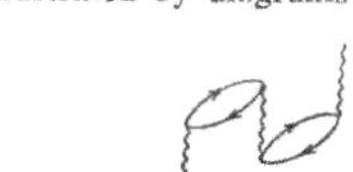

and the properties of the quanta are determined by a self-consistency condition.

Such a nuclear field theory appears to handle the delicate relationship between collective and single-particle aspects in the nuclear dynamics in a consistent manner, and the scope of the description is being intensively explored. The modification in the motion of the individual particles implied by the presence of the collective modes is expressed in terms of the renormalization of the various moments of the particles. The antisymmetry between the particles including those involved in the collective motion and, more generally, the requirements of orthogonality between the different types of quanta, are expressed in terms of interactions of exchange type. Thus, the collective modes appear just as elementary as the particle degrees of freedom out of which they are built. The various interactions between the quanta, including the width for their decay, provide the natural limitation to the description in terms of elementary modes of excitation.

The dynamic fields involved in the coupling (Eq. 1) are, of course, manifestations of the forces between the nucleons. The relationship is similar to that involved in the analysis of the average static field in the nucleus on the basis of the nucleonic interactions. These connections have proved more elusive than anticipated because of the many subtle correlation effects involved in the collective properties of the nuclear many-body system. However, the challenge that these problems present has led to extensive progress in many-body theory, which has focused attention on the deep similarities in the structure of matter in the very different forms in which it occurs in condensed atomic systems, nuclei, and elementary particles.

The nuclear vibrational fields are directly encountered in the study of the collective modes and provide a unifying quantity in terms of which one may attempt to establish the relation between the different properties of the nuclear excitations. In the efforts to determine these fields, many lines of attack are pursued including direct probing of the collective modes by nucleonic scattering, the analysis of the various observed coupling effects, and the comparison with the known static fields, as well as the deductions based on the two-nucleon forces.

Elementary Modes of Excitation

The study of the elementary modes of excitation in the nucleus and their analysis in terms of the particle-vibration coupling is an extensive domain of research, and investigations are proceeding in a variety of different directions. Thus, with the further refinement of the analysis of nuclear spectra, one expects to find many new types of collective modes, the study of which may elucidate aspects of nuclear dynamics that are at present little explored.

In another direction, the analysis of nuclei with large deformations is making contact with the study of the fission process and the related exploration of nuclear matter under the extreme conditions that are becoming accessible with the advent of high-quality heavy-ion beams. The discovery of the fission isomers and their interpretation in terms of a second minimum in the potential energy surface at very large deformations is a striking reminder of the rich structural effects that may arise from the interplay of single-particle and collective motion in the many-dimensional space of the nuclear deformations.

In yet another direction, one faces the challenge of exploring the properties of the compound nucleus in terms of the couplings between the elementary modes observed in the low energy excitation spectrum. One can envisage the development of a new branch of statistical physics concerned with the study of correlations in the structure of individual quantum states under conditions where many coupled degrees of freedom are involved and with the analysis of the concept of randomness under such conditions.

Collective Modes Involving Nucleon Transfer

While some of the collective modes, such as shape oscillations, can be described in terms of a classical picture, there are other modes associated with the more specifically quantal features of the nucleon, such as spin, isospin, and nucleon number. In recent years, there has been considerable progress in the study of collective motion with quanta carrying nucleon number. The phenomenon is related to the nuclear pairing effect, which was recognized as a systematic feature in nuclear properties at a very early stage of nuclear physics.

The striking difference in the binding energies of nuclei with even and odd numbers of nucleons finds a dramatic expression in the different fissionability properties of the even and odd isotopes of uranium. But only gradually the collective significance of the pair correlation effect and its far-reaching consequences for many nuclear properties were recognized. The correlation effect is intimately related to that of the electrons in a superconductor.

A powerful tool for the study of the nuclear pair correlations has become available in the two-nucleon transfer reactions, such as the (t,p) process, involving an incident triton (^{3}H) and an outgoing proton. In these reactions, one observes transitions of a strength much greater than would correspond to the transfer of two nucleons, each into a definite orbit in the nucleus. The enhancement implies that the pair of nucleons is transferred into a highly correlated state, and these pairs can be regarded as elementary entities, as quanta.

In some situations, one can superpose such quanta to form a vibrational spectrum, consisting of a family of states in nuclei with different numbers of neutrons or protons. In this vibrational motion, the quantity that oscillates is the amplitude of a field, the pair field, that creates two nucleons. These oscillations occur not in the usual space but in other dimensions, referred to as gauge space. This space was invented as a mathematical abstraction to express the conservation and quantization of nucleon number, or more generally of baryon number, which is identified with the angular momentum operator in gauge space. Usually, however, the nucleon number operator is assigned a rather passive role, as an overall constant of the motion, a superselection rule, that divides the phenomena into sharply separated sets, each with a definite value of nucleon number.

In the nuclear pair correlation effects, however, we have examples of phenomena that relate states with different nucleon number and that therefore involve operators, such as the orientation of the pair field in gauge space, that are complementary to the nucleon number. In the study of these nuclear phenomena, one experiences the physical significance of such operators and the reality of the new dimensions in a forceful manner.

In the development of quantal physics, the scope of symmetry concepts has greatly expanded, but also the significance of symmetry breaking has come more and more into focus. In the study of the nuclear collective modes, these themes have played a prominent role and have continued to reveal interesting new aspects, some of which may be illustrated by considering the role of isospin symmetry in the nuclear excitation spectrum.

Charge Symmetry

Crucial evidence for charge symmetry of the nuclear forces came from the original studies of mirror nuclei by Fowler, Delsasso, and Lauritsen as far back as 1936, and striking support for the more comprehensive charge independence of the forces has been provided by the impressive evidence on the occurrence of isobaric multiplets in the spectra of light nuclei, in the digestion of which the review articles emanating from the Kellogg Laboratory have played an important role.

It was generally assumed that, in heavy nuclei, the strong Coulomb forces would completely destroy the isobaric symmetry. The discovery of quite sharply defined isobaric analog states in all nuclei revealed, however, that the symmetry-breaking power of the Coulomb forces had been vastly overrated. The Coulomb interaction gives rise to major energy separations between members of a multiplet, but the relatively small variation of the Coulomb potential inside the nucleus is rather ineffective in mixing the symmetry of the wave function. Moreover, the strong tendency of the nuclear forces to favor states of low isospin acts to preserve the symmetry. One can view the isospin impurity in the nuclear wave function as a polarization effect produced by the Coulomb potential, and the relevant polarizability of the nucleus can be expressed in terms of the virtual excitation of collective monopole modes of isovector symmetry. Estimates indicate that, in the nuclear ground states, the isospin impurity does not exceed a small fraction of a percent, even in heavy nuclei.

If we consider an excitation of the nucleus, we not only can assign a total isospin quantum number, T, to the excited state, but can also talk about the isospin τ carried by the excitation. For example, if T equals the isospin T_0 of the ground state, the excitation may be isoscalar ($\tau = 0$), or isovector ($\tau = 1$), or even of higher multipolarity in isospace. However, because of the neutron excess, the generalized vacuum state for the elementary excitations (the nuclear ground state) may be highly anisotropic in isospace. The quantum number τ is therefore not in general conserved, even in the absence of Coulomb forces, corresponding to the fact that the neutrons and protons that participate in the excitation occupy different orbits. Thus, for large neutron excess, the low flying particle excitations completely violate τ symmetry. Collective modes of excitation involving neutrons and protons in different orbits also completely violate τ symmetry at the microscopic level, but the symmetry may reappear at the macroscopic level, where we consider the collective properties of the modes such as the average long wavelength density variations and the associated deformations of the average potential. In fact, it is found that the collective modes tend to preserve such a macroscopic τ symmetry, even in heavy nuclei with large neutron excess. For example, in the shape vibrations, the oscillating potential is predominantly isoscalar, while the dipole mode excited by photoabsorption is isovector to the accuracy of the available experimental data. Considerable interest attaches to the study of the small components of opposite symmetry, which can be determined, for example, by comparing the excitation of the modes by projectiles that interact differently with isoscalar and isovector fields [inelastic scattering of α particles, deuterons or individual nucleons, charge-exchange processes, such as (p,n), and other reactions].

For the isovector modes, one can go a step further and consider the relationship between the modes with T equal to $T_0 - 1$, T_0, and $T_0 + 1$ formed by the excitation of a $\tau = 1$ quantum. In the case of weak coupling between T_0 and τ, there is a simple relation between the modes, and the relative amplitudes follow from considerations of symmetry. The triplet of modes corresponds to different orientations of the isospin τ with respect to the orientation of the nuclear vacuum state in isospace and, in a nucleus with large neutron excess, the modes may have quite different properties. The weak coupling relations can then be completely violated, and it may even happen that the mode with $T_0 + 1$ is missing and reappears as a second, lower mode with $T_0 - 1$.

Occurrence of Rotational Motion

New aspects of symmetry and symmetry-breaking appear in the study of rotational motion in nuclei. The problem of whether nuclei possess rotational spectra became an issue already in the very early days of nuclear spectroscopy. Some quantal systems, such as molecules, were known to have rotational spectra, while other systems, such as atoms, do not rotate collectively.

The earliest available data on nuclear spectra, as obtained for example from the fine structure of α-decay, appeared to provide evidence against the occurrence of low-lying rotational excitations. But the discussion was hampered by the expectation that rotational motion would either be a property of all nuclei or be generally excluded, and by the assumption that the moment of inertia would have the classical value as for rigid rotation.

The establishment of the nuclear shell structure and the recognition, which came almost simultaneously, that many nuclei have equilibrium shapes deviating from spherical symmetry, created a new basis for the discussion of rotational motion in the nucleus. It was evident that the existence of a nonspherical shape implied collective rotational degrees of freedom, but one was faced with the need for a generalized treatment of rotations applicable to quantal systems that do not have a rigid or semirigid structure, such as molecules.

The existence of a deformation, taken in the general sense of an element of anisotropy in the structure of the system, may be recognized as the hallmark of quantal systems that exhibit rotational spectra. Indeed, such an element of anisotropy is required to make it possible to specify the orientation of the system. This definition of a deformation includes the lattice-like structure in molecules and rotating pieces of solid matter; a classical physical object that may seem perfectly spherical is in fact highly anisotropic, on account of the atomic constitution of matter.

The occurrence of rotational degrees of freedom may thus be said to originate in a breaking of rotational invariance. In a similar manner, the translational degrees of freedom are based upon the existence of a localized structure. However, while the different states of translational motion of a given object are related by Lorentz invariance, there is no similar invariance applying to coordinate frames rotating with respect to the matter distribution of the universe. The Coriolis and centrifugal forces that act in such coordinate frames perturb the structure of a rotating object.

In a quantal system, already the frequency of the lowest rotational excitations may be so large that the Coriolis and centrifugal forces affect the structure in a major way. The adiabatic condition, which requires these perturbations to be small, thus provides an alternative way of expressing the criterion for the occurrence of rotational spectra. It is intimately connected with the existence of an equilibrium deformation that is large compared with the zero-point fluctuations in the shape parameters.

Rotational Degrees of Freedom

The extent to which the deformation is symmetry-breaking determines the rotational degrees of freedom. A complete breaking of the rotational invariance, in the sense that the collective deformation permits a unique specification of the orientation of the system, leads to the full degrees of freedom of rotational motion in three-dimensional space, as for the asymmetric rotor, with rotational bands containing $(2I + 1)$ levels for each set of values of the total angular momentum quantum number I and its component M. A reduction in the rotational degrees of freedom occurs if the deformation preserves invariance with respect to a subgroup of rotations, including axial symmetry or a group of finite rotations or both. These operations are then part of the intrinsic degrees of freedom, and the rotational degrees of freedom are correspondingly constrained. Thus, for systems with axial symmetry, which include the well-established nuclear deformations, the bands form trajectories of states with only a single level for each IM.

The concept of rotation has gained new perspective from the recognition that the transfer of a pair of nucleons to superfluid nuclei can be viewed as rotational transitions in gauge space. The superfluid state of matter, as encountered in liquid helium, superconductors, and the majority of nuclei, is characterized by a stable deformation in gauge space; and the associated rotational motion involves a family of states with different numbers of particles, such as, for example, the ground states of nuclei with even numbers of neutrons and protons. In superconductors, the Josephson junction can be viewed in terms of rotational motion of this type involving two coupled rotors, forced to rotate with respect to each other with a frequency determined by the difference in binding energy associated with the transfer of electrons, that is, by the electrostatic potential across the junction.

One can envisage a rich variety of generalized rotational spectra associated with deformations in larger dimensions obtained by combining different spaces including isospace, gauge spaces, and orbital space. The resulting rotational band structure may involve comprehensive families of states labeled by the different quantum numbers of the internally broken symmetries, and there may be relations between quantum numbers referring to different spaces, if the deformation connects these spaces by retaining invariance with respect to combinations of different symmetry operations.

The Regge trajectories that play such a prominent role in the current study of the structure of hadrons have features reminiscent of rotational spectra, but as yet there appears to be no definite evidence concerning the degrees of freedom involved or the nature of the deformation that would define an orientation in the intrinsic structure of a hadron.

In the analysis of rotational spectra, one can develop a general phenomenological description of how the various matrix elements depend on the rotational quantum numbers, on the basis of the symmetry of the deformation and an expansion in powers of the angular momentum. A very extensive body of data on nuclear rotational spectra, to which the Norman Bridge Laboratory at the California Institute of Technology has yielded important contributions, has provided detailed tests of such a description and has thus established a firm basis for a deeper-going analysis of the dynamics of the rotational motion. Many of the observed relations can be interpreted in terms of the Coriolis coupling acting on the individual nucleons in the rotating nucleus, in a manner similar to the particle-vibration coupling described above. However, there are also indications that we have yet much to learn about the basic coupling between rotational and intrinsic motion.

The spectra of the light nuclei—and here the Kellogg Laboratory continues to be a main source of evidence—offer the opportunity to study the emergence of collective rotational motion in a system with few degrees of freedom. In such a situation, the problem of identifying the spurious particle-degrees of freedom—those out of which the rotational motion is formed—becomes especially acute.

The behavior of the rotational bands for large values of the angular momentum constitutes a frontier of vigorous activity. The current developments involve the problem of the limit of convergence of the power series expansions in the rotational frequency or angular momentum and of the many types of discontinuities that may occur in the bands for large values of the angular momentum when the rotational forces distort the nuclear structure in a major way. Some of these discontinuities may resemble phase transitions in macroscopic systems, and, in the nucleus, it may become possible to follow such transitions in terms of the properties of the individual quantum states. For the very large values of the angular momentum that can be transferred to a nucleus in an impact with a heavy ion, we encounter nuclear matter in a hitherto unknown form.

Epilogue

The development of nuclear physics in the past decades has been characterized by the great richness of the phenomena that have been encountered, and I have tried to describe a few of the concepts that have been involved in the attempt to understand these phenomena. Looking ahead, one can already see great new areas that may be explored by means of the new tools that are becoming available, and one can look forward to the inspiration which this expansion of our horizon will provide for the refinement and further development of concepts for the description of quantal phenomena (*3*).

Notes

1. For an account of C. C. Lauritsen's scientific work, see the biographical memoirs by W. A. Fowler in the *American Philosophical Society Yearbook* (1969), p. 131. The early development of nuclear research at California Institute of Technology is also vividly described by T. Lauritsen in the special issue of *Eng. Sci.* **32**, No. 9 (June 1969) published by the California Institute of Technology.
2. The development has been the result of a lively interplay of evidence and ideas coming from so many different sources that it would fall outside the scope of this presentation to attempt to mention the individual contributions. A more detailed account, with inclusion of references to the main steps in the development, has been prepared in another context, in collaboration with Ben R. Mottelson.
3. The lecture concluded by a comment on the proposed establishment of an International Science Foundation. The idea of channeling part of the resources available for scientific research through such an organization arises naturally in view of the international character of science. An International Science Foundation will be able to base its functioning on the existence of an international community of scientists who by and large share the same standards and goals for scientific research and will be in a position to evaluate research projects in an international perspective. It will be a primary task for the Foundation to help ensure that scientific talent and initiative, wherever in the world they appear, can contribute to the progress of science and, in this spirit, the Foundation might be of considerable value in promoting science in the developing countries. Plans for the establishment of an International Science Foundation were discussed at a meeting in Stockholm in July 1970, sponsored by the Royal Swedish Academy of Engineering Sciences, the Royal Swedish Academy of Sciences, Unesco, and the American Academy of Arts and Sciences, and are at present being studied by an interim committee set up at this meeting.

Aage Bohr in 1975. Courtesy of Niels Bohr Archive, Copenhagen.

ROTATIONAL MOTION IN NUCLEI

Nobel Lecture, December 11, 1975

by

AAGE BOHR
The Niels Bohr Institute and Nordita
Copenhagen, Denmark

The exploration of nuclear structure over the last quarter century has been a rich experience for those who have had the privilege to participate. As the nucleus has been subjected to more and more penetrating probes, it has continued to reveal unexpected facets and to open new perspectives. The preparation of our talks today has been an occasion for Ben Mottelson and myself to relive the excitement of this period and to recall the interplay of so many ideas and discoveries coming from the worldwide community of nuclear physicists, as well as the warmth of the personal relations that have been involved.

In this development, the study of rotational motion has had a special role. Because of the simplicity of this mode of excitation and the many quantitative relations it implies, it has been an important testing ground for many of the general ideas on nuclear dynamics. Indeed, the response to rotational motion has played a prominent role in the development of dynamical concepts ranging from celestial mechanics to the spectra of elementary particles.

EARLY IDEAS ON NUCLEAR ROTATION

The question of whether nuclei can rotate became an issue already in the very early days of nuclear spectroscopy (1, 2). Quantized rotational motion had been encountered in molecular spectra (3), but atoms provide examples of quantal systems that do not rotate collectively. The available data on nuclear excitation spectra, as obtained for example from the fine structure of a decay, appeared to provide evidence against the occurrence of low-lying rotational excitations, but the discussion was hampered by the expectation that rotational motion would either be a property of all nuclei or be generally excluded, as in atoms, and by the assumption that the moment of inertia would have the rigid-body value, as in molecular rotations. The issue, however, took a totally new form with the establishment of the nuclear shell model (4).

Just at that time, in early 1949, I came to Columbia University as a research fellow and had the good fortune of working in the stimulating atmosphere of the Pupin Laboratory where so many great discoveries were being made under the inspiring leadership of I.I. Rabi. One of the areas of great activity was the study of nuclear moments, which was playing such a crucial role in the development of the new ideas on nuclear structure.

To-day, it is difficult to fully imagine the great impact of the evidence for nuclear shell structure on the physicists brought up with the concepts of the liquid-drop and compound-nucleus models, which had provided the basis for interpreting nuclear phenomena over the previous decade (5)[1]. I would like also to recall my father's reaction to the new evidence, which presented the sort of dilemma that he would respond to as a welcome opportunity for deeper understanding. In the summer of 1949, he was in contact with John Wheeler on the continuation of their work on the fission process, and in this connection, in order to "clear his thoughts", he wrote some tentative comments on the incorporation of the contrasting evidence into a more general picture of nuclear constitution and the implications for nuclear reactions (7). These comments helped to stimulate my own thinking on the subject, which was primarily concerned with the interpretation of nuclear moments[2].

The evidence on magnetic moments, which at the time constituted one of the most extensive quantitative bodies of data on nuclear properties, presented a special challenge. The moments showed a striking correlation with the predictions of the one-particle model (9, 4), but at the same time exhibited major deviations indicative of an important missing element. The incomparable precision that had been achieved in the determination of the magnetic moments, as well as in the measurement of the hyperfine structure following the pioneering work of Rabi, Bloch, and Purcell, was even able to provide information on the distribution of magnetism inside the nucleus (10, 11).

A clue for understanding the deviations in the nuclear coupling scheme from that of the single-particle model was provided by the fact that many nuclei have quadrupole moments that are more than an order of magnitude larger than could be attributed to a single particle[3]. This finding directly implied a sharing of angular momentum with many particles, and might seem to imply a break-down of the one-particle model. However, essential features of the single-particle model could be retained by assuming that the average nuclear field in which a nucleon moves deviates from spherical symmetry (15). This picture leads to a nuclear model resembling that of a molecule, in which the nuclear core possesses vibrational and rotational degrees

[1] The struggle involved in facing up to the new evidence is vividly described by Jensen (6). Our discussions with Hans Jensen over the years concerning many of the crucial issues in the development provided for us a special challenge and inspiration.

[2] The interplay between individual-particle and collective motion was also at that time taken up by John Wheeler. Together with David Hill, he later published the extensive article on "Nuclear Constitution and the Interpretation of Fission Phenomena" (8), which has continued over the years to provide inspiration for the understanding of new features of nuclear phenomena.

[3] The first evidence for a non-spherical nuclear shape came from the observation of a quadrupole component in the hyperfine structure of optical spectra (12). The analysis showed that the electric quadrupole moments of the nuclei concerned were more than an order of magnitude greater than the maximum value that could be attributed to a single proton and suggested a deformation of the nucleus as a whole (13). The problem of the large quadrupole moments came into focus with the rapid accumulation of evidence on nuclear quadrupole moments in the years after the war and the analysis of these moments on the basis of the shell model (14).

of freedom. For the rotational motion there seemed no reason to expect the classical rigid-body value; however, the large number of nucleons participating in the deformation suggested that the rotational frequency would be small compared with those associated with the motion of the individual particles. In such a situation, one obtains definite limiting coupling schemes (see Fig. 1) which could be compared with the empirical magnetic moments and the evidence on the distribution of nuclear magnetism, with encouraging results (15, 17)[4].

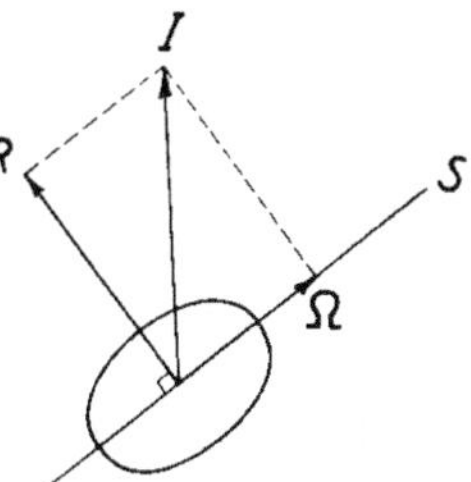

Fig. 1. Coupling scheme for particle in slowly rotating spheroidal nucleus. The intrinsic quantum number represents the projection of the particle angular momentum along the nuclear symmetry axis S, while ***R*** is the collective angular momentum of the nuclear core and is directed perpendicular to the symmetry axis, since the component along S which is a constant of the motion, vanishes in the nuclear ground state. The total angular momentum is denoted by I. The figure is from (16).

In the meantime and, in fact, at nearly the same point in space, James Rainwater had been thinking about the origin of the large nuclear quadrupole moments and conceived an idea that was to play a crucial role in the following development. He realized that a non-spherical equilibrium shape would arise as a direct consequence of single-particle motion in anisotropic orbits, when one takes into account the deformability of the nucleus as a whole, as in the liquid-drop model (19).

On my return to Copenhagen in the autumn of 1950, I took up the problem of incorporating the coupling suggested by Rainwater into a consistent dynamical system describing the motion of a particle in a deformable core. For this coupled system, the rotational motion emerges as a low-frequency component of the vibrational degrees of freedom, for sufficiently strong coupling. The rotational motion resembles a wave travelling across the nuclear surface and the moment of inertia is much smaller than for rigid rotation (see Fig. 2).

Soon, I was joined by Ben Mottelson in pursuing the consequences of the interplay of individual-particle and collective motion for the great variety of nuclear phenomena that was then coming within the range of experimental

[4] The effect on the magnetic moments of a sharing of angular momentum between the single particle and oscillations of the nuclear surface was considered at the same time by Foldy and Milford (18).

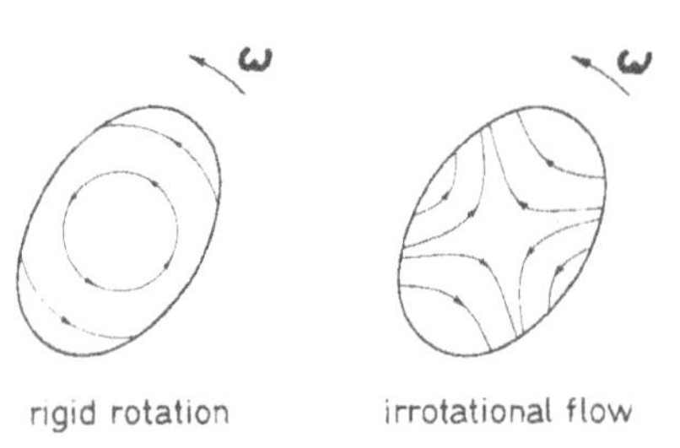

Fig. 2. Velocity fields for rotational motion. For the rotation generated by irrotational flow, the velocity is proportional to the nuclear deformation (amplitude of the travelling wave). Thus, for a spheroidal shape, the moment of inertia is $\mathcal{J} = \mathcal{J}_{rig}(\Delta R/R)^2$, where $\mathcal{J}_{rig}$ is the moment for rigid rotation, while R is the mean radius and ΔR (assumed small compared with R) is the difference between major and minor semi-axes. The figure is from (16).

studies (20). In addition to the nuclear moments, important new evidence had come from the classification of the nuclear isomers (21) and beta decay (22) as well as from the discovery of single-particle motion in nuclear reactions (23, 24). It appeared that one had a framework for bringing together most of the available evidence, but in the quantitative confrontation with experiment, one faced the uncertainty in the parameters describing the collective properties of the nucleus. It was already clear that the liquid-drop description was inadequate, and one lacked a basis for evaluating the effect of the shell structure on the collective parameters.

THE DISCOVERY OF ROTATIONAL SPECTRA

At this point, one obtained a foothold through the discovery that the coupling scheme characteristic of strongly deformed nuclei with the striking rotational band structure was in fact realized for an extensive class of nuclei. The first indication had come from the realization by Goldhaber and Sunyar that the electric quadrupole transition rates for the decay of low-lying excited states in even-even nuclei were, in some cases, much greater than could be accounted for by a single-particle transition and thus suggested a collective mode of excitation (21) . A rotational interpretation (25) yielded values for the nuclear eccentricity in promising agreement with those deduced from the spectroscopic quadrupole moments.

Soon after, the evidence began to accumulate that these excitations were part of a level sequence with angular momenta I = 0, 2, 4 . . . and energies proportional to $I\ (I+1)$ (26, 27); examples of the first such spectra are shown in Fig. 3. For ourselves, it was a thrilling experience to receive a prepublication copy of the 1953 compilation by Hollander, Perlman, and Seaborg (29) with its wealth of information on radioactive transitions, which made it possible to identify so many rotational sequences.

The exciting spring of 1953 culminated with the discovery of the Coulomb excitation process (30, 31) which opened the possibility for a systematic study of rotational excitations (30, 32). Already the very first experiments by

Huus and Zupančič (see Fig. 4) provided a decisive quantitative test of the rotational coupling scheme in an odd nucleus, involving the strong coupling between intrinsic and rotational angular momenta[5].

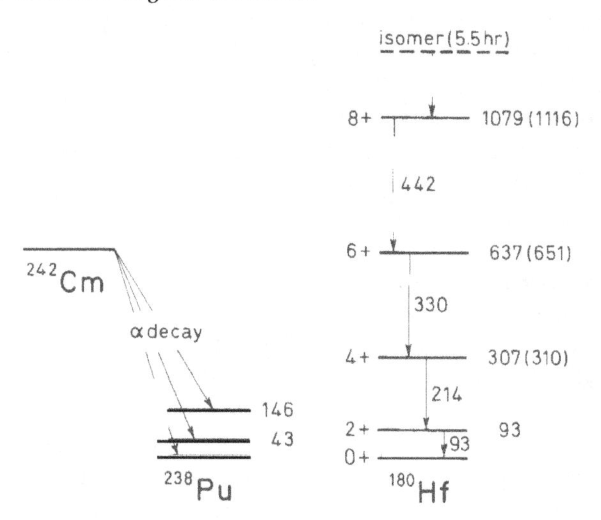

Fig. 3. Rotational spectra for ^{238}Pu and ^{180}Hf. The spectrum of ^{180}Hf (from (26)) was deduced from the observed γ lines associated with the decay of the isomeric state (28). The energies are in keV, and the numbers in parenthesis are calculated from the energy of the first excited state, assuming the energies to be proportional to $I(I+1)$.

The spectrum of ^{238}Pu was established by Asaro and Perlman (27) from measurements of the fine structure in the a decay of ^{242}Cm. Subsequent evidence showed the spin-parity sequence to be 0+, 2+ , 4+, and the energies are seen to be closely proportional to $I(I+ 1)$.

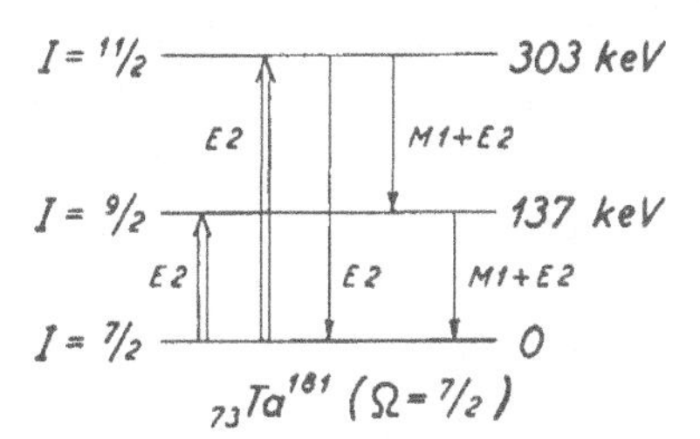

Fig. 4. Rotational excitations in ^{181}Ta observed by Coulomb excitation. In an odd-*A* nucleus with intrinsic angular momentum Ω (see Fig. 1), the rotational excitationsinvolve thesequence $I = \Omega, \Omega+1, \Omega+2,\ldots$, all with the same parity. In the Coulomb excitation process, the action of the electric field of the projectile on the nuclear quadrupole moment induces E2 (electric quadrupole) transitions and can thus populate the first two rotational excitations. The observed energies (30) are seen to be approximately proportional to $I(I+ 1)$.

The excited states decay by E2 and Ml (magnetic dipole) transitions, and the rotational interpretation implies simple intensity relations. For example, the reduced E2 matrix elements within the band are proportional to the Clebsch-Gordan coefficient $< I_i \Omega\, 20 | I_f \Omega >$, where I_i and I_f are the angular momenta of initial and final states. The figure is from (16).

[5]The quantitative interpretation of the cross sections could be based on the semi-classical theory of Coulomb excitation developed by Ter-Martirosyan (33) and Alder and Winther (34).

This was a period of almost explosive development in the power and versatility of nuclear spectroscopy, which rapidly led to a very extensive body of data on nuclear rotational spectra. The development went hand in hand with a clarification and expansion of the theoretical basis.

Fig. 5 shows the region of nuclei in which rotational band structure has so far been identified. The vertical and horizontal lines indicate neutron and proton numbers that form closed shells, and the strongly deformed nuclei are seen to occur in regions where there are many particles in unfilled shells that can contribute to the deformation.

The rotational coupling scheme could be tested not only by the sequence of spin values and regularities in the energy separations, but also by the intensity relations that govern transitions leading to different members of a rotational band (37, 38, 39). The leading order intensity rules are of a purely geometrical character depending only on the rotational quantum numbers and the multipolarity of the transitions (see the examples in Fig. 4 and Fig. 10).

The basis for the rotational coupling scheme and its predictive power were greatly strengthened by the recognition that the low-lying bands in odd-A nuclei could be associated with one-particle orbits in the deformed potential (40, 41, 42). The example in Fig. 6 shows the spectrum of ^{235}U with its high level density and apparently great complexity. However, as indicated, the states can be grouped into rotational bands that correspond uniquely to those expected from the Nilsson diagram shown in Fig. 7.

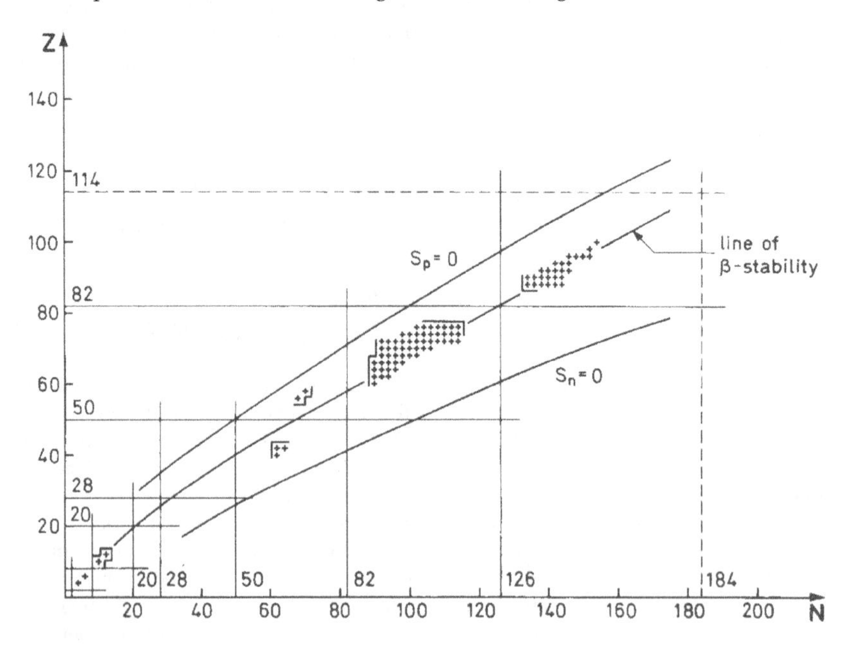

Fig. 5. Regions of deformed nuclei. The crosses represent even-even nuclei, whose excitation spectra exhibit an approximate $I(I+1)$ dependence, indicating rotational band structure. The figure is from (35) and is based on the data in (36). The curves labelled $S_n = 0$ and $S_p = 0$ are the estimated borders of instability with respect to neutron and proton emission.

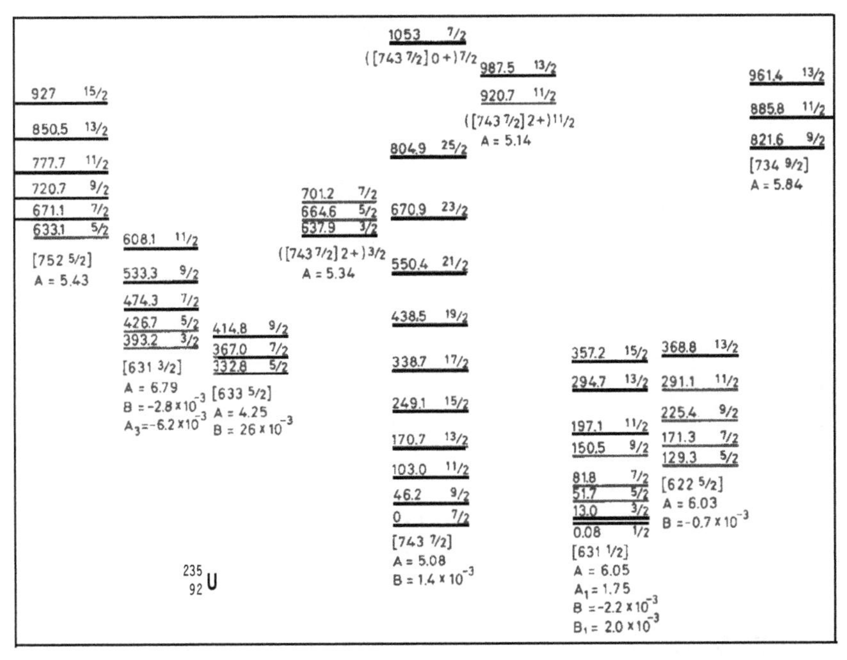

Fig. 6. Spectrum of ^{235}U. The figure is from (35) and is based on the experimental data from Coulomb excitation (43), ^{239}Pu α decay (43a), one-particle transfer (44), and the ^{234}U (n γ) reaction (45). All energies are in keV. The levels are grouped into rotational bands characterized by the spin sequence, energy dependence, and intensity rules. The energies within a band can be represented by a power series expansion of the form $E(I) = AI(I+1) + BI^2(I+1)^2 + \ldots (-1)^{I+\Omega}(I+\Omega)!\,((I-\Omega)!)^{-1}(A_{2\Omega}+B_{2\Omega}\, I(I+1) + \ldots)$, with the parameters given in the figure. The low-lying bands are labelled by the quantum numbers of the available single-particle orbits (see Fig. 7), with particle-like states drawn to the right of the ground-state band and hole-like states to the left. The bands beginning at 638, 921, and 1053 keV represent quadrupole vibrational excitations of the ground-state configuration.

The regions of deformation in Fig. 5 refer to the nuclear ground-state configurations; another dimension is associated with the possibility of excited states with equilibrium shapes quite different from those of the ground state. For example, some of the closed-shell nuclei are found to have strongly deformed excited configurations[6]. Another example of sharpe isomerism with associated rotational band structure is encountered in the metastable, very strongly deformed states that occur in heavy nuclei along the path to fission (50, 51).

[6] The fact that the first excited states in ^{16}O and ^{40}Ca have positive parity, while the low-lying single-particle excitations are restricted to negative parity, implies that these states involve the excitation of a larger number of particles. It was suggested (47) that the excited positive parity states might be associated with collective quadrupole deformations. The existence of a rotational band structure in ^{16}O was convincingly established as a result of the ^{12}C (aa) studies (48) and the observation of strongly enhanced E2-transition matrix elements (49).

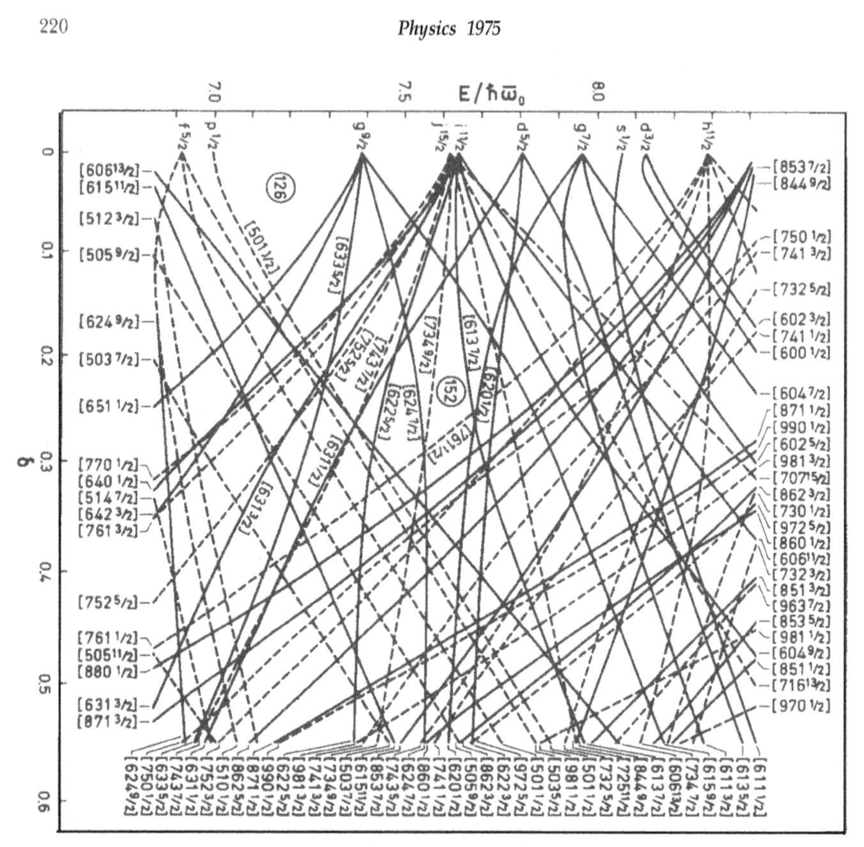

Fig. 7. Neutron orbits in prolate potential. The figure (from (35)) shows the energies of single-particle orbits calculated in an appropriate nuclear potential by Gustafson, Lamm, Nilsson, and Nilsson (46). The single-particle energies are given in units of $\hbar\bar{\omega}$, which represents the separation between major shells and, for ^{235}U, has the approximate value 6.6 MeV. The deformation parameter δ is a measure of the nuclear eccentricity; the value determined for ^{235}U, from the observed E2 transition moments, is $\delta \approx 0.25$. The single-particle states are labelled by the "asymptotic" quantum numbers $[\mathcal{N} n_3 \Lambda \Omega]$. The last quantum number Ω, which represents the component j_3, of the total angular momentum along the symmetry axis, is a constant of the motion for all values of 6. The additional quantum numbers refer to the structure of the orbits in the limit of large deformations, where they represent the total number of nodal surfaces (N), the number of nodal surfaces perpendicular to the symmetry axis (n_3), and the component of orbital angular momentum along the symmetry axis (Λ). Each orbit is doubly degenerate ($j_3 = \pm\ \Omega$), and a pairwise filling of orbits contributes no net angular momentum along the symmetry axis. For ^{235}U, with neutron number 143, it is seen that the lowest two configurations are expected to involve an odd neutron occupying the orbits [743 7/2] or [631 1/2], in agreement with the observed spectrum (see Fig. 6). It is also seen that the other observed low-lying bands in ^{235}U correspond to neighbouring orbits in the present figure.

New possibilities for studying nuclear rotational motion were opened by the discovery of marked anisotropies in the angular distribution of fission fragments (52), which could be interpreted in terms of the rotational quantum numbers labelling the individual channels through which the fissioning nucleus passes the saddle-point shape (53). Present developments in the experimental tools hold promise of providing detailed information about band structure in the fission channels and thereby on rotational motion under circumstances radically different from those studied previously.

CONNECTION BETWEEN ROTATIONAL AND SINGLE-PARTICLE MOTION

The detailed testing of the rotational coupling scheme and the successful classification of intrinsic spectra provided a firm starting point for the next step in the development, which concerned the dynamics underlying the rotational motion.

The basis for this development was the bold idea of Inglis (54) to derive the moment of inertia by simply summing the inertial effect of each particle as it is dragged around by a uniformly rotating potential (see Fig. 8). In this approach, the potential appears to be externally "cranked", and the problems concerning the self-consistent origin for the rotating potential and the limitations of such a semi-classical description have continued over the years to be hotly debated issues. The discussion has clarified many points concerning the connection between collective and single-particle motion, but the basic idea of the cranking model has stood its tests to a remarkable extent (55, 35).

The evaluation of the moments of intertia on the basis of the cranking model gave the unexpected result that, for independent-particle motion, the moment would have a value approximately corresponding to rigid rotation (56). The fact that the observed moments were appreciably smaller than the rigid-body values could be qualitatively understood from the effect of the residual interactions that tend to bind the particles into pairs with angular momentum zero. A few years later, a basis for a systematic treatment of the moment of inertia with the inclusion of the many-body correlations associated with the pairing effect was given by Migdal (57) and Belyaev (58),

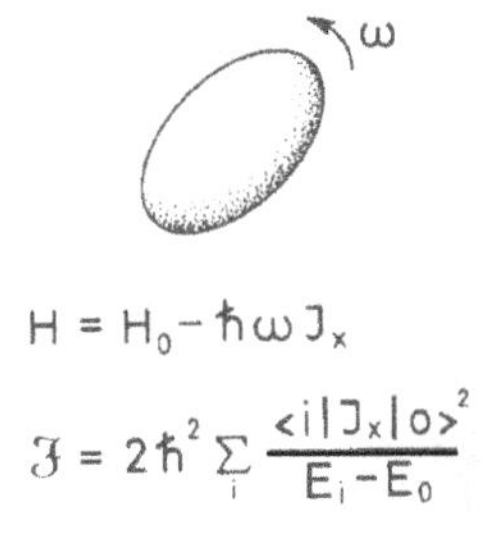

Fig. 8. Nuclear moment of inertia from cranking model. The Hamiltonian ***H*** describing particle motion in a potential rotating with frequency ω about the x axis is obtained from the Hamiltonian H_0 for motion in a fixed potential by the addition of the term proportional to the. component $\mathcal{J}_x$ of the total angular momentum, which represents the Coriolis and centrifugal forces acting in the rotating co-ordinate frame. The moment of inertia is obtained from a second-order perturbation treatment of this term and involves a sum over the excited states i. For independent-particle motion, the moment of inertia can be expressed as a sum of the contributions from the individual particles.

exploiting the new concepts that had in the meantime been developed for the treatment of electronic correlations in a superconductor (59) ; see also the following talk (60).

The nuclear moment of inertia is thus intermediate between the limiting values corresponding to rigid rotation and to the hydrodynamical picture of irrotational flow that was assumed in the early models of nuclear rotation. Indeed, the classical pictures involving a local flow provide too limited a framework for the description of nuclear rotation, since, in nuclear matter, the size of the pairs (the coherence length) is greater than the diameter of the largest existing nuclei. Macrosopic superflow of nuclear matter and quantized vortex lines may occur, however, in the interior of rotating neutron stars (61).

While these developments illuminated the many-body aspects of nuclear rotation, appropriate to systems with a very large number of nucleons, a parallel development took its starting point from the opposite side. Shell-model calculations exploiting the power of group-theoretical classification schemes and high-speed electronic computers could be extended to configurations with several particles outside of closed shells. It was quite a dramatic moment when it was realized that some of the spectra in the light nuclei that had been successfully analyzed by the shell-model approach could be given a very simple interpretation in terms of the rotational coupling scheme[7].

The recognition that rotational features can manifest themselves already in configurations with very few particles provided the background for Elliott's discovery that the rotational coupling scheme can be given a precise significance in terms of the SU_3 unitary symmetry classification, for particles moving in a harmonic oscillator potential (65). This elegant model had a great impact at the time and has continued to provide an invaluable testing ground for many ideas concerning nuclear rotation. Indeed, it has been a major inspiration to be able, even in this limiting case, to see through the entire correlation structure in the many-body wave function associated with the collective motion. Thus, for example, the model explicitly exhibits the separation between intrinsic and collective motion and implies an intrinsic excitation spectrum that differs from that of independent-particle motion in a deformed field by the removal of the "spurious" degrees of freedom that have gone into the collective spectrum.

This development also brought into focus the limitation to the concept of rotation arising from the finite number of particles in the nucleus. The rotational spectrum in the SU_3 model is of finite dimension (compact symmetry group) corresponding to the existence of a maximum angular momentum that can be obtained from a specified shell-model configuration. For low-lying bands, this. maximum angular momentum is of the order of magnitude

[7] In this connection, a special role was played by the spectrum of ^{19}F. The shell-model analysis of this three-particle configuration had been given by Elliott and Flowers (62) and the rotational interpretation was recognized by Paul (63); the approximate identity of the wave functions derived by the two approaches was established by Redlich (64).

of the number of nucleons *A* and, in some of the light nuclei, one has, in fact, obtained evidence for such a limitation in the ground-state rotational bands[8]. However, the proper place of this effect in nuclear rotations is still an open issue due to the major deviations from the schematized SU_3 picture.

GENERAL THEORY OF ROTATION

The increasing precision and richness of the spectroscopic data kept posing problems that called for a framework, in which one could clearly distinguish between the general relations characteristic of the rotational coupling scheme and the features that depend more specifically on the internal structure and the dynamics of the rotational motion[9]. For ourselves, an added incentive was provided by the challenge of presenting the theory of rotation as part of a broad view of nuclear structure. The view-points that I shall try to summarize gradually emerged in this prolonged labour (70, 71, 35).

In a general theory of rotation, symmetry plays a central role. Indeed, the very occurrence of collective rotational degrees of freedom may be said to originate in a breaking of rotational invariance, which introduces a "deformation" that makes it possible to specify an orientation of the system. Rotation represents the collective mode associated with such a spontaneous symmetry breaking (Goldstone boson) .

The full degrees of freedom associated with rotations in three-dimensional space come into play if the deformation completely breaks the rotational symmetry, thus permitting a unique specification of the orientation. If the deformation is invariant with respect to a subgroup of rotations, the corresponding elements are part of the intrinsic degrees of freedom, and the collective rotational modes of excitation are correspondingly reduced, disappearing entirely in the limit of spherical symmetry.

The symmetry of the deformation is thus reflected in the multitude of states that belong together in rotational families and the sequence of rotational quantum numbers labelling these states, in a similar manner as in the symmetry classification of molecular rotational spectra. The nuclear rotational spectra shown in Figs. 3, 4, and 6 imply a deformation with axial symmetry and invariance with respect to a rotation of 180° about an axis perpendicular to the symmetry axis (D_∞ symmetry group). It can also be inferred from the observed spectra that the deformation is invariant with respect to space and time reflection.

[8]The evidence (66, 67) concerns the behaviour of the quadrupole transition rates, which are expected to vanish with the approach to the band termination (65). This behaviour reflects the gradual alignment of the angular momenta of the particles and the associated changes in the nuclear shape that lead eventually to a state with axial symmetry with respect to the angular momentum and hence no collective radiation (68), (35).

[9]In this development, a significant role was played by the high-resolution spectroscopic studies (69) which led to the establishment of a generalized intensity relation in the E2 decay of the y-vibrational band in ^{168}Gd

The recognition of the deformation and its degree of symmetry breaking as the central element in defining rotational degress of freedom opens new perspectives for generalized rotational spectra associated with deformations in many different dimensions including spin, isospin, and gauge spaces, in addition to the geometrical space of our classical world. The resulting rotational band structure may involve comprehensive families of states labelled by the different quantum numbers of the internally broken symmetries. Relations between quantum numbers belonging to different spaces may arise from invariance of the deformation with respect to a combination of operations in the different spaces[10].

The Regge trajectories that have played a prominent role in the study of hadronic properties have features reminiscent of rotational spectra, but the symmetry and nature of possible internal deformations of hadrons remain to be established. Such deformations might be associated with boundaries for the regions of quark confinement.

The condensates in superfluid systems involve a deformation of the field that creates the condensed bosons or fermion pairs. Thus, the process of addition or removal of a correlated pair of electrons from a superconductor (as in a Josephson junction) or of a nucleon pair from a superfluid nucleus constitutes a rotational mode in the gauge space in which particle number plays the role of angular momentum (73). Such pair rotational spectra, involving families of states in different nuclei, appear as a prominent feature in the study of two-particle transfer processes (74). The gauge space is often felt as a rather abstract construction but, in the particle-transfer processes, it is experienced in a very real manner.

The relationship between the members of a rotational band manifests itself in the simple dependence of matrix elements on the rotational quantum numbers, as first encountered in the $I(I + 1)$ dependence of the energy spectra and in the leading-order intensity rules that govern transitions leading to different members of a band. The underlying deformation is expressed by the occurrence of collective transitions within the band.

For sufficiently small values of the rotational quantum numbers, the analysis of matrix elements can be based on an expansion in powers of the angular momentum. The general structure of such an expansion depends on the symmetry of the deformation and takes an especially simple form for axially symmetric systems. As an example, Fig. 9 shows the two lowest bands observed in ^{166}Er. The energies within each band have been measured with enormous precision and can be expressed as a power series that converges rather rapidly for the range of angular momentum values included in the figure. Similar expansions can be given for matrix elements of tensor operators representing

[10] A well-known example is provided by the strong-coupling fixed-source model of the pion-nucleon system, in which the intrinsic deformation is invariant with respect to simultaneous rotations in geometrical and isospin spaces resulting in a band structure with $I = T$ (72, 35).

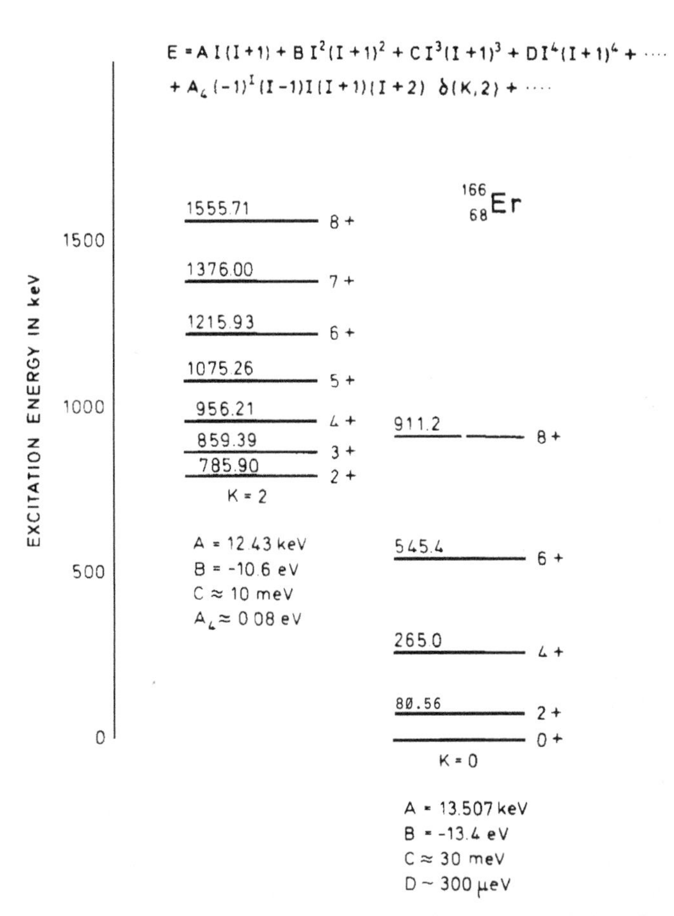

Fig. 9. Rotational bands in ^{166}Er. The figure is from (35) and is based on the experimental data by Reich and Cline (75). The bands are labelled by the component K of the total angular momentum with respect to the symmetry axis. The K = 2 band appears to represent the excitation of a mode of quadrupole vibrations involving deviations from axial symmetry in the nuclear shape.

electromagnetic transitions, β decay, particle transfer, etc. Thus, extensive measurements have been made of the E2 transitions between the two bands in ^{166}Er, and Fig. 10 shows the analysis of the empirical transition matrix elements in terms of the expansion in the angular momentum quantum numbers of initial and final states.

Such an analysis of the experimental data provides a phenomenological description of the rotational spectra in terms of a set of physically significant parameters. These parameters characterize the internal structure of the system with inclusion of the renormalization effects arising from the coupling to the rotational motion.

A systematic analysis of these parameters may be based on the ideas of the cranking model, and this approach has yielded important qualitative insight into the variety of effects associated with the rotational motion. However, in

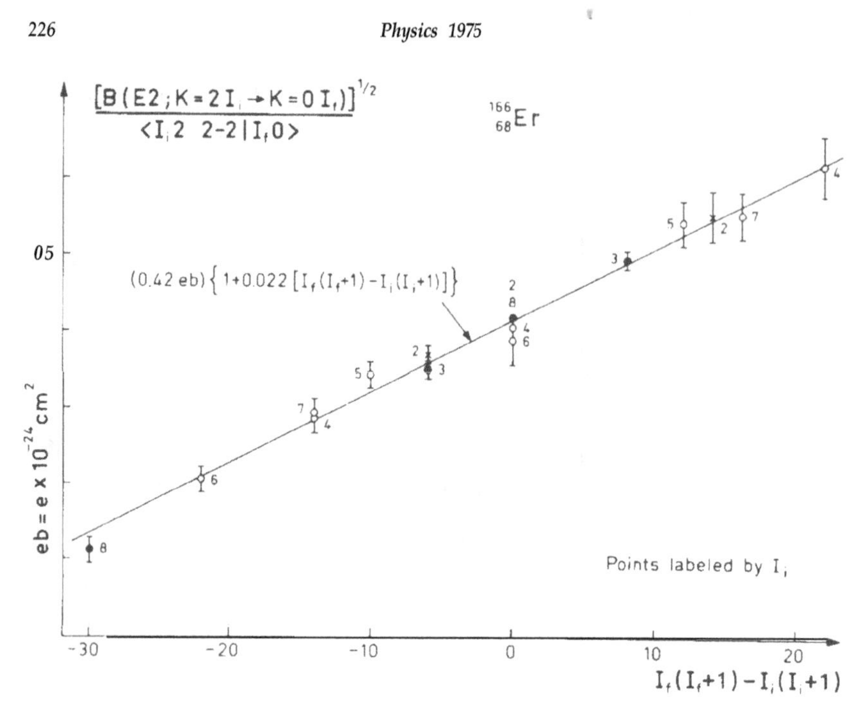

Fig. 10. Intensity relation for E2 transitions between rotational bands. The figure, which is from (35) and is based upon experimental data in (76), shows the measured reduced electric quadrupole transition probabilities B(E2) for transitions between members of the K = 2 and K = 0 bands in ^{166}Er (see Fig. 9). An expansion similar to that of the energies in Fig. 9, but taking into account the tensor properties of the E2 operator, leads to an expression for $(B(E2))^{1/2}$ which involves a Clebsch-Gordan coefficient $< I_i K_i 2 -2 | I_f K_f >$ (geometrical factor) multiplied by a power series in the angular momenta of I_i and I_f, of the initial and final states. The leading term in this expansion is a constant, and the next term is linear in $I_f(I_f+1) - I_i(I_i+1)$; the experimental data are seen to be rather well represented by these two terms.

this program, one faces significant unsolved problems. The basic coupling involved in the cranking model can be studied directly in the Coriolis coupling between rotational bands in ***odd-A*** nuclei associated with different orbits of the unpaired particle (77). The experiments have revealed, somewhat shockingly that, in many cases, this coupling is considerably smaller than the one directly experienced by the particles as a result of the nuclear rotation with respect to the distant galaxies (78). It is possible that this result may reflect an effect of the rotation on the nuclear potential itself (57, 79, 80, 35), but the problem stands as an open issue.

CURRENT PERSPECTIVES

In the years ahead, the study of nuclear rotation holds promising new perspectives. Not only are we faced with the problem already mentioned of a more deep-going probing of the rotational motion, which has become possible with the powerful modern tools of nuclear spectroscopy, but new frontiers are opening up through the possibility of studying nuclear states with very large values of the angular momentum. In reactions induced by heavy ions, it is in fact now possible to produce nuclei with as much as a hundred units of angular momentum. We thus encounter nuclear matter under quite novel conditions, where centrifugal stresses may profoundly affect the structure of the nucleus. The challenge of this new frontier has strongly excited the imagination of the nuclear physics community.

A schematic phase diagram showing energy versus angular momentum for a nucleus with mass number $A \approx 160$ is shown in Fig. 11. The lower curve representing the smallest energy, for given angular momentum, is referred to as the yrast line. The upper curve gives the fission barrier, as a function of angular momentum, estimated on the basis of the liquid-drop model (81). For $I \approx 100$, the nucleus is expected to become unstable with respect to fission, and the available data on cross sections for compound-nucleus formation in heavy ion collisions seem to confirm the approximate validity of this estimate of the limiting angular momentum (82).

Present information on nuclear spectra is confined almost exclusively to a small region in the left-hand corner of the phase diagram, and a vast extension of the field is therefore coming within range of exploration. Special interest attaches to the region just above the yrast line, where the nucleus, though highly excited, remains cold, since almost the entire excitation energy is concentrated in a single degree of freedom. One thus expects an excitation spectrum with a level density and a degree of order similar to that near the ground state. The extension of nuclear spectroscopy into this region may therefore offer the opportunity for a penetrating exploration of how the nuclear structure responds to the increasing angular momentum.

In recent years, it has been possible to identify quantal states in the yrast region up to $I \approx 20$ to 25, and striking new phenomena have been observed.

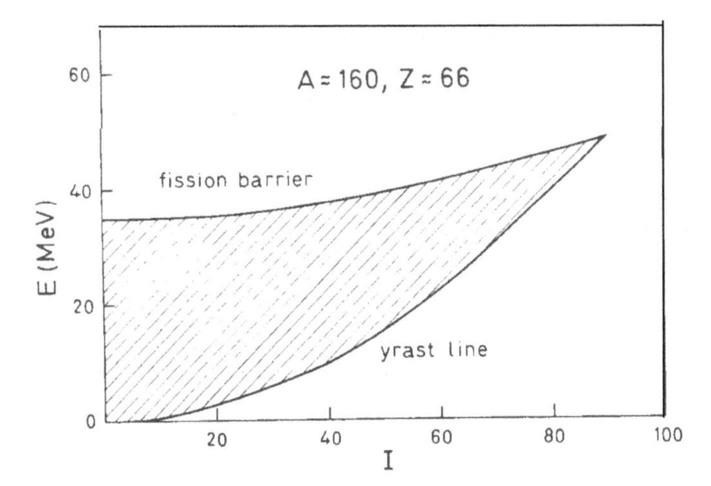

Fig. 11. Nuclear phase diagram for excitation energy versus angular momentum. The yrast line and the fission barrier represent estimates, due to Cohen, Plasil, and Swiatecki (81), based on the liquid-drop model, with the assumption of the rigid-body value for the moment of inertia.

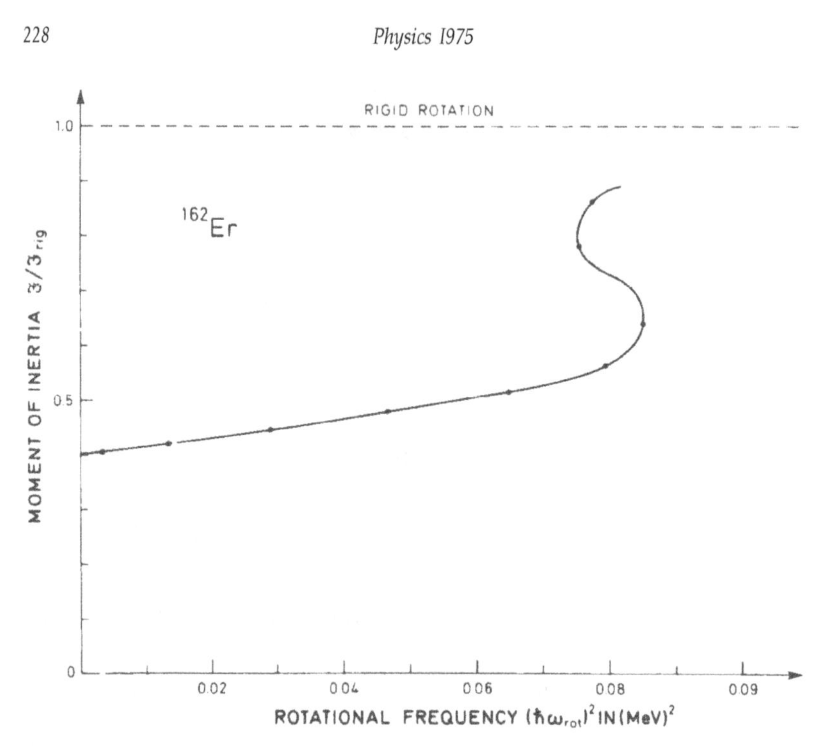

Fig. 12. Moment of inertia as function of rotational frequency. The figure is from (83) and is based on the experimental data of Johnson, Ryde, and Hjorth (84). The rotational frequency is defined as the derivative of the rotational energy with respect to the angular momentum and is obtained by a linear interpolation in the variable $I(I+1)$ between the quantal states. The moment of inertia is defined in the usual manner as the ratio between the angular momentum and the rotational frequency.

An example is shown in Fig. 12, in which the moment of inertia is plotted against the rotational frequency. This "back-bending"effect was discovered here in Stockholm at the Research Institute for Atomic Physics, and has been found to be a rather general phenomenon.

In the region of angular momenta concerned, one is approaching the phase transition from superfluid to normal nuclear matter, which is expected to occur when the increase in rotational energy implied by the smaller moment of inertia of the superfluid phase upsets the gain in correlation energy (85). The transition is quite analogous to the destruction of superconductivity by a magnetic field and is expected to be associated with an approach of the moment of inertia to the rigid-body value characteristic of the normal phase.

The back-bending effect appears to be a manifestation of a band crossing, by which a new band with a larger moment of inertia and correspondingly smaller rotational frequency for given angular momentum, moves onto the yrast line. Such a band crossing may arise in connection with the phase transition, since the excitation energy for a quasiparticle in the rotating potential may vanish, even though the order parameter (the binding energy of the correlated pairs) remains finite, in rather close analogy to the situation in gapless superconductors (86). In fact, in the rotating potential, the angular

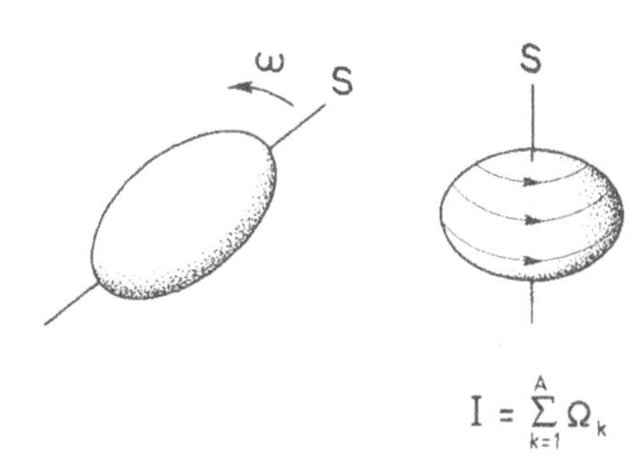

Fig. 13. Collective rotation contrasted with alignment of particle angular momenta along a symmetry axis.

momentum carried by the quasiparticle tends to become aligned in the direction of the axis of rotation. The excitation of the quasiparticle is thus associated with a reduction in the angular momentum and, hence, of the energy that is carried by the collective rotation (87).

It must be emphasized that, as yet, there is no quantitative interpretation of the striking new phenomena, as exemplified by Fig. 12. One is facing the challenge of analyzing a phase transition in terms of the individual quantal states.

For still larger values of the angular momentum, the centrifugal stresses are expected to produce major changes in the nuclear shape, until finally the system becomes unstable with respect to fission. The path that a given nucleus follows in deformation space will depend on the interplay of quantal effects associated with the shell structure and classical centrifugal effects similar to those in a rotating liquid drop. A richness of phenomena can be envisaged, but I shall mention only one of the intriguing possibilities.

The classical centrifugal effects tend to drive the rotating system into a shape that is oblate with respect to the axis of rotation, as is the case for the rotating earth. An oblate nucleus, with its angular momentum along the symmetry axis, will represent a form for rotation that is entirely different from that encountered in the low-energy spectrum, where the axis of rotation is perpendicular to the symmetry axis (see Fig. 13). For a nucleus spinning about its symmetry axis, the average density and potential are static, and the total angular momentum is the sum of the quantized contributions from the individual particles. In this special situation, we are therefore no longer dealing with a collective rotational motion characterized by enhanced radiative transitions, and the possibility arises of yrast states with relatively long lifetimes (88). If such high-spin metastable states (super-dizzy nuclei) do in fact occur, the study of their decay will provide quite new opportunities for exploring rotational motion in the nucleus at very high angular momenta.

Thus, the study of nuclear rotation has continued over the years to be alive and to reveal new, challenging dimensions. Yet, this is only a very special aspect of the broader field of nuclear dynamics that will be the subject of the following talk.

REFERENCES

1. Thibaud, J., Comptes rendus 191, 656 (1930)
2. Teller, E., and Wheeler, J. A., Phys. Rev. 53, 778 (1938)
3. Bjerrum, N. in *Nernst Festschrift* , p. 90, Knapp, Halle 1912
4. Mayer, M. G., Phys. Rev. 75, 209 (1949) ; Haxel, O., Jensen, J. H. D., and Suess, H. E., Phys. Rev. 75, 1766 (1949)
5. Bohr, N., Nature 137, 344 (1936) ; Bohr, N., and Kalckar, F., Mat. Fys. Medd. Dan. Vid. Selsk. 14, no, 10 (1937); Weisskopf, V. F., Phys. Rev. 52, 295 (1937) ; Meitner, L., and Frich, 0. R., Nature 243, 239 (1939); Bohr, N., and Wheeler, J. A., Phys. Rev. 56, 426 (1939); Frenkel, J., J. Phys. (USSR) *1,* 125 (1939); see also Phys. Rev. 55, 987 (1939)
6. Jensen, J. H. D., Les Prix Nobel en 1963, p. 153 Imprimerie Royale P. A. Norstedt & Shiner, Stockholm 1964
7. Bohr, *N., Tentative Comments on Atomic and Nuclear Constitution.* Manuscript dated August 1949. Niels Bohr Archives, The Niels Bohr Institute, Copenhagen
8. Hill, D. L., and Wheeler, J. A., Phys. Rev. 89, 1102 (1953)
9. Schmidt, T., Z Physik 106, 358 (1937)
10. Bitter, F., Phys. Rev. 75, 1326 (1949); 76, 150, (1949)
11. Bohr, A., and Weisskopf, V. F., Phys. Rev. 77, 94 (1950)
12. Schüler, H., and Schmidt, Th., Z. Physik 94, 457 (1935)
13. Casimir, H. B. *G., On the Interaction Between Atomic Nuclei and Electrons,* Prize Essay, Taylor's Tweede Genootschap, Haarlem (1936); see also Kopfermann, H., Kernmomente, Akademische Verlagsgesellschaft, Leipzig 1940
14. Townes, C. H., Foley, H. M., and Low, W., Phys. Rev. 76, 1415 (1949)
15. Bohr, A., Phys. Rev. 81, 134 (1951)
16. Bohr, A., *Rotational States of Atomic Nuclei,* Munksgård, Copenhagen 1954
17. Bohr, A., Phys. Rev. 81, 331 (1951)
18. Foldy, L. L., and Milford, F. J., Phys. Rev. 80, 751 (1950)
19. Rainwater, J., Phys. Rev. 79, 432 (1950)
20. Bohr, A., and Mottelson, B. R., Mat. Fys. Medd. Dan. Vid. Sselsk. 27, no. 16 (1953)
21. Goldhaber, M., and Sunyar. A. W., Phys. Rev. 83, 906 (1951)
22. Mayer, M. G., Moszkowski, S. A., and Nordheim, L. W., Rev. Mod. Phys. 23, 315 (1951) Nordheim, L. W., Rev. Mod. Phys. 23, 322 (1951)
23. Barschall, H. H., Phys. Rev. 86, 431 (1952)
24. Weisskopf, V. F., Physica 18, 1083 (1952)
25. Bohr, A., and Mottelson, B. R., Phys. Rev. 89, 316 (1953)
26. Bohr, A., and Mottelson, B. R., Phys. Rev. 90, 717 (1953)
27. Asaro, F., and Perlman, I., Phys. Rev. 91, 763 (1953)
28. Burson, S. B., Blair, K. W., Keller, H. B., and Wexler, S., Phys. Rev. 83, 62 (1951)
29. Hollander, J. M., Perlman, I., and Seaborg, G. T., Rev. Mod. Phys. 25, 469 (1953)
30. Huus, T. and Zupančič, C., Mat. Fys. Medd. Dan. Vid. Selsk. 28, no. 1 (1953)
31. McClelland, C. L., and Goodman, C., Phys. Rev. 91, 760 (1953)
32. Heydenburg, N. P., and Temmer, G. M., Phys. Rev. 94, 1399 (1954) and 100, 150 (1955)
33. Ter-Martirosyan, K. A., Zh. Eksper. Teor. Fiz. 22, 284 (1952)
34. Alder, K., and Winther, A. Phys. Rev. 91, 1578 (1953)
35. Bohr, A., and Mottelson, B. R., *Nuclear Structure,* Vol. II. Benjamin, W. A. Inc., Reading, Mass. 1975
36. Sakai, M., Nuclear Data Tables *A8,* 323 (1970) and *A10,* 511 (1972)
37. Alaga, G., Alder, K., Bohr, A., and Mottelson, B. R., Mat. Fys. Medd. Dan. Vid. Selsk. 29, no. 9 (1955)
38. Bohr, A., Froman, P. O., and Mottelson, B. R., Mat. Fys. Medd. Dan. Vid. Selsk. 29, no. 10 (1955)

39. Satchler, G. R., Phys. Rev. 97, 1416 (1955)
40. Nilsson, S. G., Mat. Fys. Medd. Dan. Vid. selsk. 29, no. 16 (1955)
41. Mottelson, B. R., and Nilsson, S. G., Phys. Rev. 99, 1615 (1955)
42. Gottfried, K.,.Phys. Rev. 203, 1017 (1956)
43. Stephens, F. S., Holtz, M. D., Diamond, R. M., and Newton, J. O., Nuclear Phys. *A115,* 129 (1968)
43a. Cline, J. E., Nuclear Phys. ***A106,*** 481 (1968)
44. Elze, Th. W., and Huizenga, J. R., Nuclear Phys, ***A133,*** 10 (1969); Braid. T. H., Chasman, R. R., Erskine. J. R., and Friedman, A. M., Phys. Rev. C1, 275 (1970)
45. Jurney, E. T., *Neutron Capture Gamma-Ray Spectroscopy,* Proceedings of the International Symposium held in Studsvik, p. 431, International Atomic Energy Agency, Vienna 1969
46. Gustafson, C., Lamm. I. L., Nilsson, B., and Nilsson, S. G., Arkiv Fysik 36, 613 (1967)
47. Morinaga, H., Phys. Rev. ***101, 254*** (1956)
48. Carter, E. B., Mitchell, G. E., and Davies, R. H., Phys. Rev. 133, B1421 (1964)
49. Gorodetzky, S., Mennrath, P., Benenson, W., Chevallier, P., and Scheibling, F., J. phys. radium 24, 887 (1963)
50. Polikanov, S. M., Druin, V. A., Karnaukhov, V. A., Mikheev, V. L., Pleve, A. A., Skobelev, N. K., Subbotin, V. G., Ter-Akop'yan, G. M., and Fomichev, V. A., J. Exptl. Theoret. Phys. (USSR) 42, 1464 (1962); transl. Soviet Physics JETP ***15,*** 1016
51. Specht, H. J., Weber, J., Konecny, E., and Heunemann, D., Phys. Letters 41B, 43 (1972)
52. Winhold, E. J., Demos, P. T., and Halpern, I., Phys. Rev. 87, 1139 (1952)
53. Bohr, A. in ***Proc.*** Intern. *Conf. on the Peaceful Uses of Atomic Energy,* vol. 2, p. 151, United Nations, New York, 1956
54. Inglis, D. R., Phys. Rev. 96, 1059 (1954)
55. Thouless, D. J., and Valatin, J. G., Nuclear Phys. 31, 211 (1962)
56. Bohr, A., and Mottelson, B. R., Mat. Fys. Medd. Dan. Vid. Selsk. 30, no. 1 (1955)
57. Migdal, A. B., Nuclear Phys. ***13, 655*** (1959)
58. Belyaev, S. T., Mat. Fys. Medd. Dan. Selsk. 31, no. 11 (1959)
59. Bardeen, J., Cooper, L. N., and Schrieffer, J. R.. Phys. Rev. 106, 162 and 108, 1175 (1957)
60. Mottelson. B. R., this volume, p. 82.
61. Ruderman, M.. Ann. Rev. Astron. and Astrophys. 10, 427 (1972)
62. Elliott, J. P., and Flowers, B. H., Proc. Roy. Soc. (London) ***A229, 536,*** (1955)
63. Paul, E. B., Phil. Mag. 2,311 (1957)
64. Redlich, M. G., Phys. Rev. ***110,468*** (1958)
65. Elliott, J. P., Proc. Roy. Soc. (London) ***A245,*** 128 and 562 (1958)
66. Jackson, K. P., Ram, K. B., Lawson, P. G., Chapman N. G., and Allen, K. W., Phys. Letters 30B, 162 (1969)
67. Alexander, T. K., Häusser, O., McDonald, A. B., Ferguson, A. J., Diamond, W. T., and Litherland, A. E., Nuclear Phys. ***A179, 477*** (1972)
68. Bohr, A. in *Intern. Nuclear Phys. Conf.* , p. 489, Ed.-in-Chief R. L. Becker, Academic Press, New York 1967
69. Hansen, P. G., Nielsen, 0. B., and Sheline, R. K., Nuclear Phys. 12, 389 (1959)
70. Bohr, A., and Mottelson, B. R., Atomnaya Energiya ***14,*** 41 (1963)
71. Bohr, A. in *Symmetry Properties of* Nuclei, Proc. of 15th Solvay Conf. on Phys. 1970, p. 187. Gordon and Breach Science Publ., London 1974
72. Henley, E. M., and Thirring, W., Elementary Quantum *Field Theory* , McGraw Hill, New York 1962
73. Anderson, P. W., Rev. Mod. Phys. 38, 298 (1966)
74. Middleton, R., and Pullen, D. J., Nuclear Phys. 52, 77 (1964); see also Broglia, R.

A., Hansen, O., and Riedel, C. *Advances in Nuclear Physics* 6, 287, Plenum Press, New York 1973

75. Reich, C. W., and Cline, J. E., Nuclear Phys. ***A159,*** 181 (1970)
76. Gallagher, C. J., Jr., Nielsen, 0. B., and Sunyar, A. W., Phys. Letters 16, 298 (1965); Günther, C., and Parsignault, D. R., Phys. Rev. 153, 1297 (1967); Domingos, J. M., Symons, G. D., and Douglas, A. C., Nuclear Phys. ***A180, 600*** (1972)
77. Kerman, A. K., Mat. Fys. Medd. Dan. Vid. Selsk. 30, no. 15 (1956)
78. Stephens, F. (1960), quoted by Hyde, E., Perlman, I., and Seaborg, G. T. in ***The*** Nuclear ***Properties*** of ***the Heavy Elements,*** Vol. II, p. 732, Prentice Hall, Englewood Cliffs, N.J. 1964; Hjorth, S. A., Ryde, H., Hagemann, K. A., Løvhøiden, G., and Waddington, J. C., Nuclear Phys. ***A144,*** 513 (1970) ; see also the discussion in *(35)*
79. Belyaev, S. T., Nuclear Phys. 24, 322 (1961)
80. Hamamoto, I., Nuclear Phys. ***A232, 445, (*** 1974)
81. Cohen, S., Plasil, F., and Swiatecki, W. J., Ann, Phys. 82, 557 (1974)
82. Britt, H. C., Erkilla, B. H., Stokes, R. H., Gutbrod, H. H., Plasil, F., Ferguson, R. L., and Blann, M., Phys. Rev. C in press; Gauvin, H., Guerrau, D., Le Beyec, Y., Lefort, M., Plasil, F., and Tarrago, X., Phys. Letters, 58B, 163 (1975)
83. Bohr, A., and, Mottelson, B. R., ***The Many Facets*** of ***Nuclear*** Structure. Ann. Rev. Nuclear Science 23, 363 (1973)
84. Johnson, A., Ryde, H., and Hjorth, S. A., Nuclear Phys. ***A179, 753*** (1972)
85. Mottelson, B. R., and Valatin, J. G., Phys. Rev. Letters 5, 511 (1960)
86. Goswami, A., Lin, L., and Struble, G. L., Phys. Letters 25B, 451 (1967)
87. Sephens, F. S., and Simon, R. S., Nuclear Phys. ***Al83, 257*** (1972)
88. Bohr, A., and Mottelson, B. R., Physica Scripta ***10A,*** 13 (1974)

SOCIETÀ ITALIANA DI FISICA

SCUOLA INTERNAZIONALE DI FISICA « E. FERMI »

LXIX CORSO - VARENNA SUL LAGO DI COMO - VILLA MONASTERO - 26 Luglio - 9 Agosto 1976

Conference photograph of the 1976 International School of Physics "Enrico Fermi", Course LXIX, in Villa Monastero, Varenna, on Lake Como, Italy (26 July–9 August). Reproduced with kind permission of Società Italiana di Fisica from Proceedings of the International School of Physics "Enrico Fermi", Course LXIX, Elementary Modes of Excitation in Nuclei, Eds. A. Bohr and R. A. Broglia, North-Holland, Amsterdam (1977), copyright©1976, by Società Italiana di Fisica.

1. G. A. Wolzak
2. B. A. Brown
3. N. Haik
4. R. Wadsworth
5. A. Pittino
6. J. W. Smits
7. F. Auger
8. A. Poves
9. P. Bonche
10. A. M. Saruis
11. R. Satchler
12. B. H. Wildenthal
13. B. R. Mottelson
14. I. Talmi
15. A. Bohr
16. R. A. Broglia
17. A. Kerman
18. A. Winther
19. D. R. Bès
20. A. Brondi
21. A. Covello
22. C. Guet
23. K. F. Liu
24. C. Giusti
25. J. E. Koops
26. L. Pires de Brito
27. E. M. Pereira da Silva
28. I. Hamamoto
29. O. Akyüz
30. W. Alberico
31. K. Halkia
32. P. F. Bortignon
33. I. M. Bortignon
34. S. Shlomo
35. J. Almberger
36. S. Åberg
37. G. Andersson
38. R. Leonardi
39. C. Marville
40. R. Janssens
41. C. Michel
42. J. Vervier
43. G. Poggi
44. T. Koppel
45. S. Turck
46. T. R. Halemane
47. J. P. Blaizot
48. G. Auger
49. M. Poljsak
50. J. Dudek
51. G. Giberti
52. M. Morando
53. E. Mazzi
54. E. Flerackers
55. J. De Raedt
56. L. Richter
57. M. Waroguier
58. B. Berthier
59. G. Venden Berghe
60. A. Molinari
61. J. Poitou
62. H. Hussein
63. G. Ripka
64. J. Tepel
65. S. Islam
66. J. S. Lilley
67. S. Stringari
68. O. Hansen
69. D. Proetel
70. F. Stephens
71. H. Bokemeyer
72. A. J. Kreiner
73. M. R. Maier
74. A. Insolia
75. A. Stefanini
76. M. Baldo
77. G. Lo Bianco
78. R. Cenni
79. C. Dasso
80. K. Ogawa
81. O. Civitarese
82. V. Krishnamurthy
83. H. Müller
84. V. Manfredi
85. D. Evers

Some Aspects of Rotational Motion.

A. Bohr

The Niels Bohr Institute and Nordita - Copenhagen

Rotational motion in nuclei is a very extensive subject. Indeed, this mode of excitation, characterized by its many simple quantitative relations, is among the most intensively studied. The present lectures will be confined to a discussion of a few selected topics, in relation to those dealt with in other lectures and seminars (*).

1. – Rotational degrees of freedom. Deformation and its symmetry.

Not all quantal systems possess rotational spectra. A structural feature is required that can define an orientation of the system as a whole. The occurrence of such a « deformation » may be viewed as a spontaneous breaking of rotational symmetry, and the rotational mode is the corresponding « Goldstone boson ».

The rotational degrees of freedom depend on the symmetry of the deformation. The full degrees of freedom associated with rotations in three-dimensional space come into play if the deformation completely breaks rotational symmetry, thus permitting a unique specification of the orientation of the system. The orientation is described by three angular variables, such as the Euler angles $\omega = (\phi\theta\psi)$; three conjugate angular momenta and three quantum numbers are needed to specify the rotational motion. Two of these are IM, the total angular momentum and its component on a fixed z-axis. To obtain a third, one considers the components I_1, I_2, I_3 of $\boldsymbol{I}$ along a set of intrinsic or body-fixed axes. These components are invariant with respect to rotations of the fixed, external axes and thus commute with the components I_x, I_y, I_z. The eigenvalues of $I_{1,2,3}$ are the same as for $I_{x,y,z}$ and the eigenvalue of I_3 is conventionally denoted by K. Thus, a complete set of rotational wave functions can be specified by IMK. These are the $\mathscr{D}^I_{MK}(\omega)$ functions, also referred to as the rotation matrices (*i.e.* the representations of the rotation group).

(*) A more comprehensive discussion including references can be found in [1].

For a completely asymmetric system, the rotational spectrum associated with any given intrinsic state thus contains $2I+1$ states for each set of values of IM.

If the deformation is invariant under some rotations, it is no longer possible to specify the orientation of the system uniquely, and the rotational degrees of freedom are correspondingly reduced. The rotations that leave the deformation invariant become part of the intrinsic degrees of freedom. For the nuclear deformations, the most important symmetries are axial symmetry and invariance with respect to a rotation of 180° about an axis perpendicular to the symmetry axis ($\mathscr{R}$-symmetry).

For a deformation with axial symmetry, the component of intrinsic angular momentum along the axis of symmetry is a constant of the motion K. There is no way of specifying a collective orientation with respect to this axis and hence no collective rotations about the axis. The rotational degrees of freedom have the constraint $I_3 = K$, and the spectrum comprises only a single state IM, for $I \geqslant |K|$.

The $\mathscr{R}$-invariance applies, for example, to a deformation of spheroidal (or ellipsoidal) shape. For such a symmetry, the operation $\mathscr{R}$ is part of the intrinsic degrees of freedom ($\mathscr{R}_i$), and the rotational motion is constrained so that the corresponding rotation of the system as a whole ($\mathscr{R}_e$) is equal to the intrinsic operation

$$\mathscr{R}_e = \mathscr{R}_i \tag{1}$$

in analogy to the constraint $I_3 = K$, resulting from axial symmetry. The constraint (1) reduces the number of states in the rotational spectrum by a factor of two.

The combination of axial symmetry with $\mathscr{R}$-invariance leads to an invariance group for the deformation that is non-Abelian (since the operators $\mathscr{R}$ and I_3 do not commute). The representations of this group (D_∞) are for most part two-dimensional, corresponding to the degeneracy between intrinsic states with opposite sign of K.

In the special case of $K = 0$, there is no degeneracy, and the intrinsic state $K = 0$ is an eigenstate of $\mathscr{R}$, with eigenvalue $r = \pm 1$. Since the rotational wave function for $K = 0$ is $Y_{IM}(\theta\phi)$, the operation $\mathscr{R}_e$, which inverts the direction of the axis, has the value $(-1)^I$, and the constraint (1) implies

$$(-1)^I = r\,. \tag{2}$$

For $K \neq 0$, the operation $\mathscr{R}$ connects the degenerate states with opposite K, and the total state satisfying the constraint (1) becomes a definite linear combination of states with $\pm K$. The two degenerate intrinsic states together give only a single rotational state IM with $I \geqslant |K|$.

These well-known results have been discussed without explicit reference to the structure of the total wave functions, in order to emphasize that they follow directly from considerations of symmetry. They can be extended to deformations that are invariant with respect to any subgroup of the full rotation group. One may also consider a combination of rotation with other symmetry operations, such as space and time reflection. For example, an axially symmetric deformation with an octupole component (pear shape) is invariant with respect to $\mathscr{S} = \mathscr{R}\mathscr{P}$, where $\mathscr{P}$ is space reflection, but not with respect to $\mathscr{R}$ and $\mathscr{P}$ separately. The constraint (2) is now replaced by

$$\left\{\begin{aligned} &(-1)^I \pi = s\,, \\ &I\pi = \begin{cases} 0+,\,1-,\,2+\,, & s=+1\,, \\ 0-,\,1+,\,2-\,, & s=-1\,, \end{cases} \end{aligned}\right. \tag{3}$$

for the case of $K = 0$. The eigenvalue of $\mathscr{S}$ is denoted by s.

Deformations may also occur in other dimensions, such as isospace or the gauge space associated with the pairing. One then obtains rotational spectra extending into these spaces, where isospin and particle number play the role of angular momenta. (See, for example, the discussion of pair rotations in [2].)

2. – General form of rotational matrix elements.

The intimate relation between states that belong to the same rotational family manifests itself in the dependence of matrix elements on the rotational quantum numbers. This dependence reflects the symmetry of the deformation and the tensorial structure of the operators. For not too large values of the angular momentum, the general form of these matrix elements can be obtained by exploiting the approximate separation between intrinsic and rotational motion.

The condition for separability can be expressed as the adiabatic requirement that the rotational frequencies be small compared with the intrinsic ones; the internal structure will then follow the rotational motion almost adiabatically. This condition is intimately connected with the existence of a well-defined deformation, which requires that the zero-point fluctuations in the deformation parameters be small compared with the equilibrium values of these parameters ($\Delta\alpha \ll \alpha$). This can be seen for simple models, but appears to be a very general relation, connected with the fact that, for $\Delta\alpha \sim \alpha$, the rotational and internal degrees of freedom merge. (Of course, the system may have special internal excitations with very small frequencies, in which case it may be necessary to consider explicitly the coupling between rotation and these near-degenerate internal excitations.)

Separation of the motion implies that the total wave function, when expressed in suitably chosen co-ordinates, is a product of intrinsic and rotational parts

$$\Psi(q,\omega) = \Phi_{\text{intr}}(q)\,\varphi_{\text{rot}}(\omega)\,. \tag{4}$$

The intrinsic and collective co-ordinates q and ω (and their conjugate momenta) are functions of the co-ordinates, momenta and spin values of the particles, but, at this stage, we need not be concerned with the explicit relationship.

For a nucleus with axial symmetry and $\mathscr{R}$-invariance, the wave functions (4) take the form, for $K=0$,

$$\left\{\begin{aligned} &\Psi_{r,K=0,IM} = \left(\frac{2I+1}{8\pi^2}\right)^{\frac{1}{2}} \Phi_{r,K=0}(q)\,\mathscr{D}^I_{M0}(\omega)\,, \\ &(-1)^I = r\,. \end{aligned}\right. \tag{5}$$

For $K \neq 0$, the wave function satisfying the constraint (1) can be written

$$\left\{\begin{aligned} \Psi_{KIM} &= \frac{1}{\sqrt{2}}(1+\mathscr{R}_i^{-1}\mathscr{R}_e)\left(\frac{2I+1}{8\pi^2}\right)^{\frac{1}{2}} \Phi_K(q)\,\mathscr{D}^I_{MK}(\omega) = \\ &= \left(\frac{2I+1}{16\pi^2}\right)^{\frac{1}{2}} \left(\Phi_K(q)\,\mathscr{D}^I_{MK}(\omega) + (-1)^{I+K}\Phi_{\bar{K}}(q)\,\mathscr{D}^I_{M-K}(\omega)\right), \\ I &= K, K+1, \ldots \qquad\qquad (K>0) \end{aligned}\right. \tag{6}$$

with the notation and relations

$$\left\{\begin{aligned} &\Phi_{\bar{K}} = \mathscr{R}_i^{-1}\Phi_K = \mathscr{T}\Phi_K\,, \\ &\mathscr{R}_e\,\mathscr{D}^I_{MK} = \exp[-i\pi I_2]\,\mathscr{D}^I_{MK} = (-1)^{I+K}\mathscr{D}^I_{M-K} \qquad (\mathscr{R} = \mathscr{R}_2(\pi))\,, \end{aligned}\right. \tag{7}$$

where $\mathscr{T}$ denotes time reversal. Axes perpendicular to the symmetry axis are equivalent; following the conventional notation, we have chosen $\mathscr{R}$ to be a rotation through 180° about the 2-axis. (The phase between the two parts of the wave functions (6) depends on the phase convention for $\Phi_{\bar{K}}$ and the choice of $\mathscr{R}$, but, irrespective of these choices, involves the factor $(-1)^I$.)

The existence of two terms in the states (6) with a relative phase that alternates in sign for successive values of I implies that one may also view the band as consisting of two families (or trajectories) each with $\Delta I = 2$ (as for the $K=0$ bands) and distinguished by the value of the signature

$$\sigma \equiv (-1)^{I+K}\,. \tag{8}$$

The two families with different σ may be associated with the two intrinsic states, as will be discussed below.

The above wave functions represent the symmetry structure of the bands. However, the separated form is only a first approximation. Coriolis and centrifugal forces acting in the rotating (or body-fixed) frame give rise to coupling between intrinsic and rotational motion, corresponding to a total Hamiltonian

$$H = H_{\text{intr}} + H_{\text{rot}} + H_{\text{coupl}} , \tag{9}$$

where the coupling term depends on rotational as well as intrinsic variables. The coupling gives rise to wave functions that are superpositions of states of the form (5) or (6) (band mixing). However, we can also think of these couplings as giving rise to renormalized, effective operators that act in the unperturbed basis. Formally, this corresponds to the performance of a canonical transformation $\exp[iS]$, and the matrix elements of an operator F can therefore be obtained by evaluating the transformed or renormalized operator $\exp[iS]F\exp[-iS]$ in the unperturbed basis. For not too large values of the angular momentum, the transformation S and the renormalized operators may be expressed as a power series expansion in the rotational angular momentum.

To illustrate the general procedure, we consider as a simple example the rotational energy. Since the energy is a scalar, it cannot depend on the orientation angles ω, and only involves the intrinsic components of the angular momentum $I_{1,2,3}$, in addition to the internal variables qp.

For a band with $K = 0$, the effective energy operator is diagonal with respect to I_3 and can only depend on the combination $I_1^2 + I_2^2$:

$$(H_{\text{rot}})_{\Delta K=0} = F(qp, I_1^2 + I_2^2) . \tag{10}$$

Expanding in powers of $I_1^2 + I_2^2$, which has the value $I(I+1)$ (since $I_3 = 0$), we obtain the familiar form

$$E_{\text{rot}} = AI(I+1) + BI^2(I+1)^2 , \tag{11}$$

where $A, B, \ldots$ are expectation values for the intrinsic state.

For $K \neq 0$, we obtain additional signature-dependent terms in the rotational energy arising from the part of the energy operator acting between the parts of the wave function with $I_3 = \pm K$. This contribution to the rotational Hamiltonian has $\Delta K = \pm 2K$ and is of the general form

$$\begin{cases} (H_{\text{rot}})_{\Delta K=\pm 2K} = (I_-)^{2K} F(qp, I_1^2 + I_2^2) + \text{Herm. conj.} , \\ I_- = I_1 - iI_2 . \end{cases} \tag{12}$$

The resulting addition to the rotational energy can be expressed in the form

$$\Delta E_{rot} = (-1)^{I+K} \frac{(I+K)!}{(I-K)!} (A_{2K} + B_{2K} I(I+1) + \ldots) . \tag{13}$$

(Since K^2 is a constant for the band, one can expand either in powers of $I(I+1)$ or of $I(I+1)-K^2$.)

For tensor operators, such as a multipole moment $\mathscr{M}(\lambda\mu)$, the expansion is obtained by a transformation to the intrinsic co-ordinate system

$$\mathscr{M}(\lambda\mu) = \sum_{\nu} \mathscr{M}(\lambda\nu; qpI_{\pm}) \mathscr{D}^{\lambda}_{\mu\nu}(\omega) . \tag{14}$$

The effective intrinsic moments are scalars (independent of the orientation ω) and are conveniently expanded in powers of the intrinsic angular-momentum components $I_{\pm} = I_1 \pm iI_2$.

We shall not here go into the explicit form of the resulting matrix elements. The main point to be emphasized is that one obtains model-independent

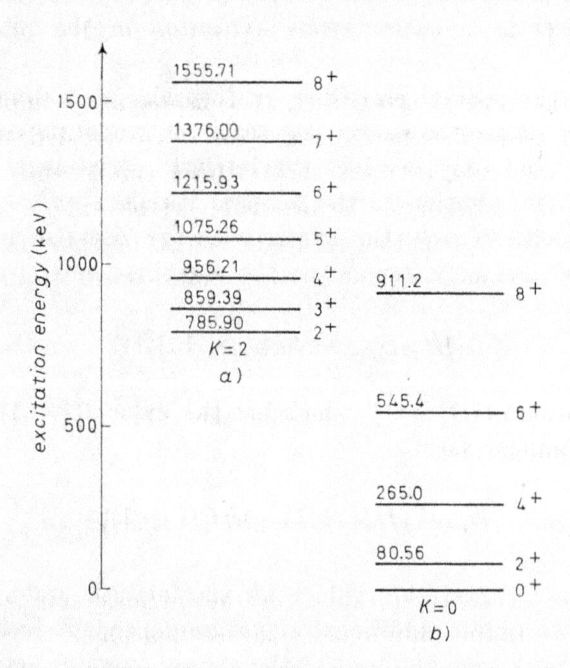

Fig. 1. – Rotational bands in ^{166}Er. The figure is from [1] and is based on the experimental data by REICH and CLINE [3]. The $K=2$ band appears to represent the excitation of a γ-vibration (quadrupole mode with ellipsoidal deformations about the spheroidal equilibrium). The energies can be represented by the expression $E = AI(I+1) + BI^2(I+1)^2 + CI^3(I+1)^3 + DI^4(I+1)^4 + A_4(-1)^I(I-1)I(I+1)(I+2)\delta(K,2)$ with the parameters $a)$ $A = 12.43$ keV, $B = -10.6$ eV, $C \approx 10$ meV, $A_4 = 0.08$ eV; $b)$ $A = 13.507$ keV, $B = -13.4$ eV, $C \approx 30$ meV, $D \sim 300$ μeV.

forms of the expansion based only on the symmetry of the deformation and the tensorial structure of the operators. These expressions provide an analysis of the experimental data in terms of a set of physically significant parameters.

An example of such an analysis is given in fig. 1, which shows the two lowest bands observed in ^{166}Er. The energies within each band have been measured with very great precision and can be expressed as a power series that converges rather rapidly for the range of angular momenta included in the figure.

Extensive measurements have also been made of the $E2$ transitions between the two bands in ^{166}Er, and their analysis is illustrated in fig. 2. To leading

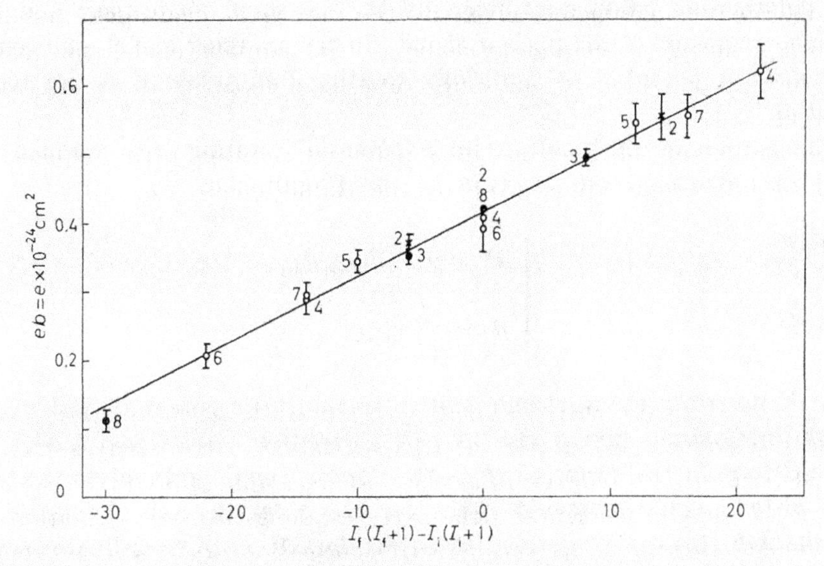

Fig. 2. – Intensity relation for $E2$ transitions between rotational bands with $\Delta K = 2$. The figure, which is from [1] and is based on experimental data in [4], shows the measured reduced electric-quadrupole transition probabilities $B(E2)$ for transitions between members of the $K = 2$ and $K = 0$ bands in ^{166}Er (see fig. 1). The quantity plotted is $(B(E2; K = 2\, I_i \to K = 0\, I_f))^{\frac{1}{2}} \langle I_i\, 2\, 2 - 2 | I_f\, 0\rangle$, and the straight line that approximately fits the data is given by $0.42eb\{1 + 0.022\,[I_f(I_f + 1) - I_i(I_i + 1)]\}$. The points are labelled by the spin I_i of the initial state.

order, the transition operator is given by (14) with $\mathscr{M}(\lambda\nu)$ independent of $I_\pm$ and leads to a reduced matrix element proportional to the vector addition coefficient $\langle I_i\, K_i = 2\; 2\; -2 | I_f\, K_f = 0\rangle$. To next order, one obtains contributions from terms in $\mathscr{M}(\lambda, \nu = \pm 1)$ linear in $I_\pm$, and these give matrix elements containing the additional factor $I_f(I_f + 1) - I_i(I_i + 1)$.

As an alternative to the expansion in powers of the angular momentum, one may expand the effective operators in powers of the rotational frequency. Such an expansion of the rotational energy is found to converge more rapidly than the expansion in powers of $I(I + 1)$, for reasons that are not well understood at the present time.

3. – Cranking model.

The interpretation of the parameters in the general expansion of the rotational properties requires a more detailed insight into the dynamics of the rotational motion. A basis for such an interpretation is provided by the cranking model.

The existence of an average deformed potential in the intrinsic system is very substantially documented, especially by the successful interpretation of the low-lying bands in odd-A nuclei in terms of one-particle motion in a potential with deformation parameters given by the measured quadrupole and higher multipole moments of the nuclear shape. In the cranking model, one considers the deformed potential as uniformly rotating and treats it as an external potential.

The motion of the nucleons in a potential rotating with frequency ω_{rot} about the intrinsic 1-axis is given by the Hamiltonian

$$(15)\qquad \begin{cases} H' = H - \hbar\omega_{\text{rot}} J_1 \,, \\[2mm] J_1 = \sum\limits_{k=1}^{A} (j_1)_k \,, \end{cases}$$

where H describes the nucleonic motion in the static potential (and may also include interactions giving rise to pair correlation, vibrational modes, etc.). The last term in (15) (referred to as the Coriolis coupling) involves the components of the angular momenta of the particles along the axis of rotation. The Hamiltonian (15) can be derived by a transformation to co-ordinates referring to the rotating system. One may also note that J_1 is the operator generating a rotation of the co-ordinate system. Thus, the difference between the time derivatives of an operator F evaluated in the fixed and rotating system is equal to

$$(16)\qquad \left(\frac{\mathrm{d}F}{\mathrm{d}t}\right)' - \left(\frac{\mathrm{d}F}{\mathrm{d}t}\right) = -i\omega_{\text{rot}}[J_1, F] = \frac{i}{\hbar}[H' - H, F] \,.$$

From (15), one obtains the properties of the rotating nucleus as a function of the rotational frequency. For example, the energy (the expectation value of H), to second order in ω_{rot}, is given by

$$(17)\qquad \begin{cases} E_{\text{rot}} = \dfrac{1}{2}\mathscr{I}\,\omega_{\text{rot}}^2 \,, \\[3mm] \mathscr{I} \;\;= 2\hbar^2 \sum\limits_i \dfrac{\langle i|J_1|0\rangle^2}{E_i - E_0} \,, \end{cases}$$

where 0 denotes the unperturbed state (assumed to be nondegenerate, as in a $K = 0$ band), while i labels the excited states of H.

The expansion in frequency can be converted into an expansion in I by evaluating the angular momentum as a function of ω_{rot}. Thus, to first order,

$$I = \langle J_1 \rangle = \mathscr{I}\omega_{rot} . \tag{18}$$

Hence, the energy for rotation about the 1-axis is

$$E_{rot} = \frac{\hbar^2}{2\mathscr{I}} I_1^2 , \tag{19}$$

and we can, therefore, identify the parameter A in the quantal expression (11) with the coefficient in (19)

$$A = \frac{\hbar^2}{2\mathscr{I}} . \tag{20}$$

One can proceed in this manner to evaluate all the different terms in the power series expansion in I discussed above. Thus, although the cranking model involves a semi-classical approximation only valid for large I (since the axis of rotation—the direction of I—is considered as a fixed quantity), one has a unique way of interpreting the results in terms of the general quantal expressions derived on the basis of symmetry considerations.

An evaluation of the moment of inertia (17), for independent-particle motion, yields a value close to the classical expression for the rotation of a rigid body. The rigid-body moment is obtained as an exact result in certain limiting situations, such as for particles in a harmonic-oscillator potential with the equilibrium deformation appropriate to the configuration considered and in the Thomas-Fermi approximation. The occurrence of the rigid-body moment of inertia can be understood from the fact that the velocity distribution in the deformed static potential tends to be isotropic at each point, and that this isotropy is not affected by the Coriolis or centrifugal forces. When there is no net flow in the intrinsic system, one sees from a fixed system the velocity $\boldsymbol{\omega} \times \boldsymbol{r}$, as for rigid rotation. In general, the evaluation of the moment of inertia (17) for independent-particle motion yields values that exhibit fluctuations about an average corresponding to the rigid-body value; there is as yet no detailed understanding of these fluctuations.

The empirical moments of inertia are smaller than the rigid-body values, typically by a factor 2 to 3. This effect appears to be associated with the pair correlations, which lead to an increase in the energy denominators in (17), since the excitations involve a breaking of pairs, and to a decrease in the numerators, connected with the tendency of the interactions to bind pairs into states with angular momentum zero.

The pair correlations imply superfluid behaviour of nuclear matter, and the rotational motion is intermediate between the limiting values corresponding to rigid rotation and irrotational flow. However, the concept of a local flow provides too narrow a framework for the description of nuclear rotation since the pairing is a highly nonlocal phenomenon. (In nuclear matter, the size of pair (the coherence length ξ) is large compared with the diameter of the largest existing nuclei ($\xi \sim \hbar v_F/\Delta \sim R(\hbar\omega_0/\Delta)$), where $\hbar\omega_0$ is the spacing between major shells.)

Though the cranking model has provided qualitative insight into a variety of phenomena associated with the nuclear rotation, the quantitative analysis has encountered serious difficulties. The basic interaction can be studied directly in the coupling between rotational bands in odd-A nuclei, and is found to be significantly smaller than the Coriolis coupling $\omega_{\text{rot}} j_1$. This observation may reflect an effect of the rotation on the nuclear potential itself. Such additional terms proportional to the rotational frequency are indeed to be expected from velocity-dependent interactions, in particular from the pairing force. These problems will be dealt with in [5].

The cranking model operates with a uniformly rotating potential. However, the frequency of rotation is not an exact constant of the motion; for given I, there will be a spread of frequencies, but, if many degrees of freedom are contributing to the rotation, one expects relatively small fluctuations about the average. In certain situations, however, a treatment in terms of a single fixed rotational frequency is inadequate. Thus, we shall later encounter examples of a weak coupling between near-lying or crossing bands with two somewhat different frequencies, where a coupled-channel treatment is required. Quite generally, when a single degree of freedom contributes significantly to the rotational motion, there will be departures from uniform rotation, and it becomes necessary to treat explicitly the coupling of this degree of freedom to the rotational motion.

An example is the treatment of an odd-A nucleus in terms of the last odd particle coupled to the core, the latter being described by appropriate collective degrees of freedom. Thus, if the core can be treated as an axially symmetric rotor with a fixed moment of inertia, such a particle-rotor model takes the form

$$H = H_{\text{rotor}} + H_{\text{particle}} = \frac{\hbar^2}{2\mathcal{J}_0}(\boldsymbol{R})^2 + T + V\,, \tag{21}$$

where $\boldsymbol{R}$ is the angular momentum of the core assumed perpendicular to the symmetry axis ($R_3 = 0$), $\mathcal{J}_0$ its moment of inertia, T the kinetic energy of the particle, and V its potential energy, which depends on co-ordinates relative to the rotor. With the substitution $\boldsymbol{R} = \boldsymbol{I} - \boldsymbol{j}$, the Hamiltonian can be written

$$\left\{\begin{aligned} H &= H_0 + H_c\,,\\ H_0 &= T + V + \frac{\hbar^2}{2\mathcal{J}_0}(\boldsymbol{I})^2,\\ H_c &= -\frac{\hbar^2}{2\mathcal{J}_0}(j_1 I_1 + j_2 I_2) + \frac{\hbar^2}{2\mathcal{J}_0}(j_1^2 + j_2^2 - j_3^2)\,. \end{aligned}\right. \tag{22}$$

The first term H_0 describes the coupled system in the adiabatic approximation and has eigenstates of the form (6), assuming axial symmetry and $\mathscr{R}$ invariance for the rotor. The second term H_c gives the particle-rotation coupling and also includes the « recoil » term; for example, a second-order treatment of the particle-rotation coupling gives the additional energy

$$\delta E_{\text{rot}} = -\frac{\hbar^2}{2\mathscr{I}_0^2}\,\delta\mathscr{I}\, I(I+1)\,, \tag{23}$$

where $\delta\mathscr{I}$ is the contribution of the particle to the moment of inertia, as given by the cranking model (see (17)). Thus, for $\delta\mathscr{I} \ll \mathscr{I}_0$, we obtain a rotational energy with a total moment of inertia $\mathscr{I} = \mathscr{I}_0 + \delta\mathscr{I}$. To higher order in $\delta\mathscr{I}$, however, one gets deviations from the cranking model. If many particles are added to the rotor, the coupling between the particles via the recoil term ensures that each particle is coupled to a rotor consisting of the core plus the remaining particles, and one regains the results of the cranking model, provided $\delta\mathscr{I}$ for each particle is small compared with the total moment of inertia.

The cranking model operates with an externally cranked potential. One can study how the rotational mode develops self-consistently by proceeding in analogy to the analysis of the vibrational modes, as discussed in [6]. To the particle-vibration coupling associated with a deformation of the nuclear potential corresponds the particle-rotation coupling

$$H_{\text{coupl}} = -\alpha \frac{\partial V}{\partial \varphi_1}\,, \tag{24}$$

where α is the angular amplitude of the collective rotational motion and φ_1 the azimuth with respect to the axis of rotation. The coupling (24) added to the Hamiltonian H for motion in the static deformed field restores rotational invariance of the total Hamiltonian, to leading order; as a consequence of this invariance, a normal-modes treatment (RPA) of the coupling (24) gives a zero-frequency collective mode representing rotational motion. Since the field $\partial V/\partial\varphi_1$ generates excitations with $K\pi = 1+$ (when acting on a $K\pi = 0+$ ground state), the rotational mode is seen in this way to develop out of the $K\pi = 1+$ particle excitations. Such a treatment also yields explicit expressions for the collective rotational co-ordinates ω in terms of the variables associated with the particle excitations, in analogy to the analysis of vibrations based on the particle-vibration coupling. The approach further substantiates the basis for the cranking model; in particular, the mass parameter for the zero-frequency mode equals the expression (17) for the moment of inertia. (The normal-modes treatment of the rotation, in terms of the coupling (24), can be viewed as a special case of the self-consistent treatment [7], in which the interactions merely generate a rotation of the static potential.)

4. – Stability of highly spinning nuclei.

New frontiers in the study of nuclear rotational motion have been opened through the possibility of studying nuclear states with very large values of the angular momentum. To see the scope of this new field, we may first ask how much angular momentum a nucleus can hold before it disintegrates. With increasing rotation, the nucleus becomes more and more unstable with respect to fission and finally breaks into two parts. This instability can be explored in terms of the liquid-drop model, which has been an important guide in the understanding of fission phenomena. The equilibrium shapes and fission barriers for rotating charged liquid drops have been especially studied by COHEN, PLASIL and SWIATECKI [8].

For a rotating liquid drop, the equilibrium shape as a function of angular momentum is found by minimizing the energy

$$\mathscr{E}(\alpha) = \mathscr{E}_{\text{surf}} + \mathscr{E}_{\text{Coul}} + \mathscr{E}_{\text{rot}} \tag{25}$$

considered as a function of the shape parameters α. The rotational energy is

$$\mathscr{E}_{\text{rot}} = \frac{\hbar^2 I^2}{2\mathscr{I}(\alpha)}, \tag{26}$$

where the angular momentum I is taken to be along the axis with the largest moment of inertia $\mathscr{I}(\alpha)$, for given deformation. We shall assume $\mathscr{I}$ to have the rigid-body value, since, as discussed below, the pair correlations are expected to vanish for large values of the angular momentum.

When the drop is set into rotation, it takes the shape of an oblate spheroid. In fact, such a deformation leads to an increase in the moment of inertia

$$\mathscr{I}_{\text{obl}} \approx \mathscr{I}_0\left(1 + \frac{2}{3}\delta\right), \qquad \delta = \frac{\Delta R}{R}, \tag{27}$$

where $\mathscr{I}_0$ is the moment for spherical shape. Hence, the equilibrium corresponds to a deformation

$$\delta \approx \frac{2}{3}\frac{1}{C}\frac{\hbar^2}{2\mathscr{I}_0} I^2, \tag{28}$$

where C is the restoring force arising from surface and Coulomb terms.

For sufficiently fast rotation, also the axially symmetric equilibrium configuration becomes unstable. For this distortion, the increase in $\mathscr{I}$ is only of second order in the deformation parameter (γ), and the rotational energy thus gives a reduction in the restoring force, proportional to I^2. A finite rotational frequency is therefore required for this instability, which takes the drop into an ellipsoidal shape. Finally, the rotating drop becomes unstable with respect to fission.

If we first ignore Coulomb effects, the dependence of the critical angular momentum on A follows from a dimensional argument

$$\frac{\mathscr{E}_{\text{rot}}}{\mathscr{E}_{\text{surf}}} \sim \frac{I^2}{A^{\frac{7}{3}}}, \qquad I_{\text{crit}} \sim A^{\frac{7}{6}}, \tag{29}$$

since the moment of inertia is proportional to $A^{5/3}$ and the surface energy to $A^{2/3}$. Coulomb effects lower the fission barrier and thus lead to a decrease in I_{crit}, which is well known to vanish for a critical value of Z^2/A (≈ 45), for which the nonrotating drop becomes unstable with respect to quadrupole deformations.

Figure 3 shows the liquid-drop estimate for $I_{\text{crit}}(A)$, for nuclei along the β-stability line. The dashed curve indicates the transition to triaxial shapes.

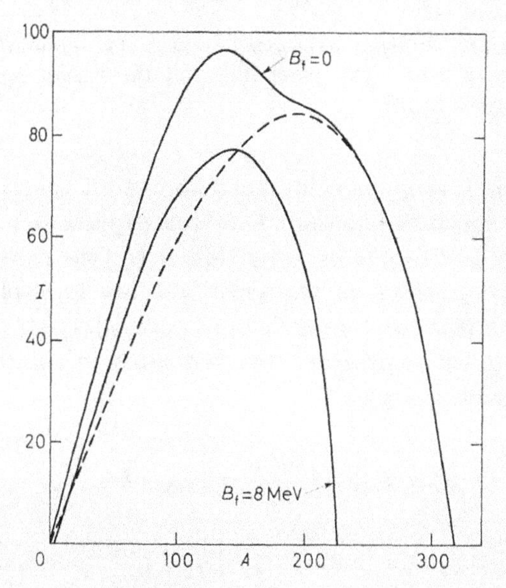

Fig. 3. – Stability limits for rotating liquid drop. The figure is based on ref. [8] and shows, for nuclei along the β-stability line, the values of the angular momentum I for which the fission barrier vanishes ($B_f = 0$) and has a value of 8 MeV (corresponding to an average nucleonic separation energy).

Information on the magnitude of angular momentum that a nucleus can hold without instantly fissioning comes from the measured cross-sections for the fusion of two colliding heavy ions. The evidence seems to be consistent with limiting values approximately equal to those deduced from the liquid-drop model (see, for example, [9]).

For a given nucleus, we can exhibit the region of stability in an E-I phase diagram (see fig. 4), which shows the yrast line and the fission barrier, as estimated from the liquid-drop model, for a nucleus with $A \approx 160$. Special interest attaches to the region just above the yrast line, where the nucleus,

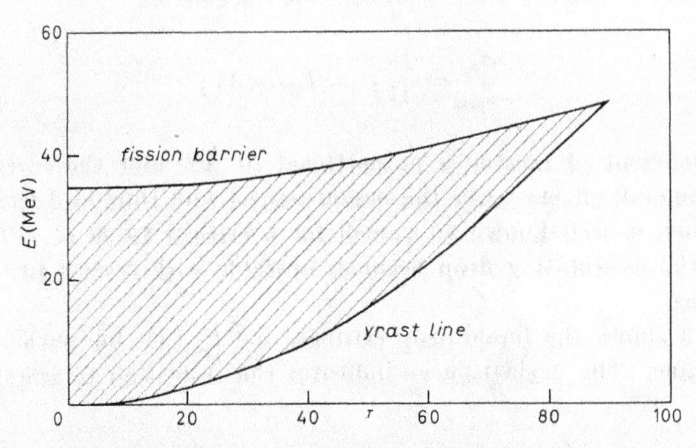

Fig. 4. – Phase diagram showing excitation energy *vs.* angular momentum, for a nucleus with $A \approx 160$, $Z \approx 66$. The yrast line and the fission barrier are based on the liquid-drop estimates in [8].

though highly excited, remains cold, since almost all the excitation energy is concentrated in the rotational motion. One thus expects an excitation spectrum with a level spacing and degree of order similar to that near the ground state.

The highly excited states on the yrast line are unstable with respect to particle emission (nucleons and α-particles in particular). If the particle carries away l units of angular momentum, the reduction in rotational energy gives an effective separation energy

$$S_l = S_{l=0} - l\,\frac{\partial E_{\text{rot}}}{\partial I} = S_{l=0} - \hbar\omega_{\text{rot}}\, l\,, \tag{30}$$

which becomes negative for sufficiently large l. The nucleons in a nucleus with large probability carry values of l up to a maximum value of order $A^{1/3}$, and the critical value of ω_{rot} leading to $S_l = 0$ for a neutron or proton is thus proportional to $A^{-1/3}$. Since the moment of inertia increases as $A^{5/3}$, the corresponding critical value of I is of order $A^{4/3}$ and hence increases faster with A than the critical value for instability with respect to fission (see (29)). However, since $S_{l=0}$ is only a small fraction of the Fermi kinetic energy, the numerical coefficient is small ($I_{\text{crit}} \approx 0.1 A^{4/3}$) and, hence, particle emission dominates over fission instability in lighter nuclei (see fig. 5). In addition to nucleon ejection, the emission of α-particles may be an important decay channel; this process is favoured by the relatively small binding energies and the large l

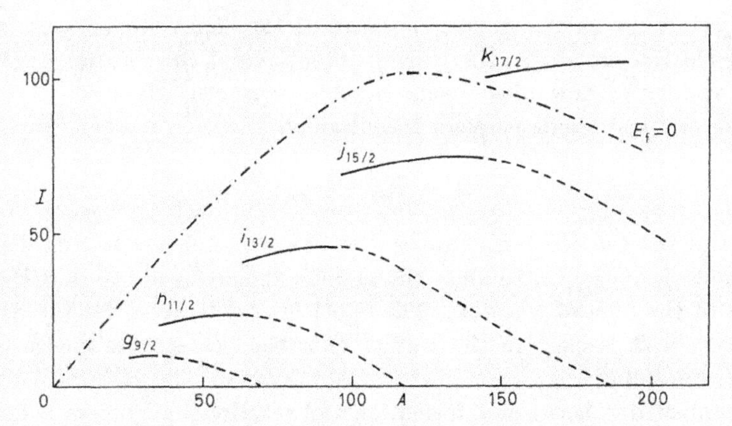

Fig. 5. – Unbound neutron levels at the Fermi surface. The figure shows the values of I for which the neutron levels $g_{9/2}$, $h_{11/2}$, ... become occupied on the yrast line. Full-drawn lines refer to unbound levels, dotted lines to bound levels. The Fermi energy ε_F is taken to be -8 MeV and the binding energies refer to a Woods-Saxon potential with an oblate deformation of $\delta = -0.3$. (Since the orbits considered are strongly concentrated along the equator, the binding energies are expected to depend mainly on the average radii perpendicular to the axis of rotation and thus for any given deformation can be approximately obtained from the values in the figure by an appropriate shift in the value of A.) Particle instability in rotating nuclei has been considered by H. C. CHANG, T. DÖSSING, S. FRAUENDORF and H. SCHULTZ (private communications).

values that the α-particles can carry, and in many situations is expected to be the limiting factor for stability with respect to particle emission.

As mentioned above, the rotating liquid drop takes on an oblate shape with eccentricity given by (28). For a nucleus with $A \approx 160$, the restoring force has a value of about 80 MeV (using the surface energy in the semi-empirical mass formula), and one obtains

$$\delta \approx 6 \cdot 10^{-5} I^2 \approx \begin{cases} 0.15\,, & I = 50\,, \\ 0.3\;\,, & I = 70\,. \end{cases} \tag{31}$$

Comparable deformation effects are associated with the nuclear shell structure, and the nuclear shapes as a function of I therefore result from a rather delicate competition between the macroscopic centrifugal effects incorporated in the liquid-drop model and the quantal effects associated with the shell structure. Current studies by several groups are exploring the succession of equilibrium shapes as a function of I for different nuclei (see, for example, [10] and [11]).

5. – Nucleonic motion in rapidly rotating potentials. Rotational alignment.

The basis for the analysis of shell structure effects in high-spin nuclei and of the yrast spectra is the study of single-particle motion in rotating potentials.

The subject is an extensive one, because of the many dimensions involved, including the nuclear shape, the pairing potential and the rotational frequency. A great variety of new phenomena can be envisaged.

In the cranked single-particle Hamiltonian

$$H' = H_{sp} - \hbar\omega_{rot} j_1 \tag{32}$$

the effect of the Coriolis term can be viewed as a tendency to establish orbits with definite $j_1 = \Omega_1$, *i.e.* to align the angular momenta of the particles in the direction of the axis of rotation. The effect is counteracted by the azimuthal deformation with respect to the 1-axis. (For the present, we ignore pair correlations; see below.)

The competition between deformation and rotational alignment is illustrated in fig. 6 for particle motion in a harmonic-oscillator potential. The figure

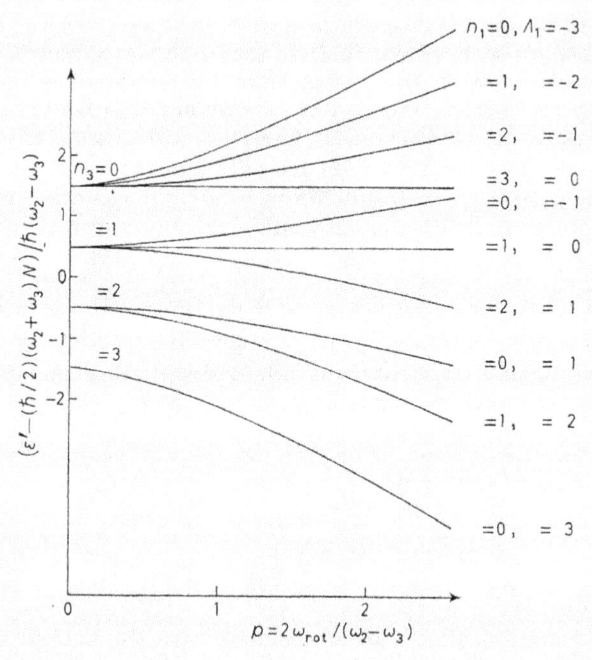

Fig. 6. – Eigenvalues for particle motion in rotating harmonic-oscillator potential. The figure shows the eigenvalues ε' of the cranked Hamiltonian $H' = H_{sp} - \hbar\omega_{rot} l_1$ as a function of ω_{rot}, for particles in the $N = n_1 + n_2 + n_3 = 3$ shell for a prolate nucleus ($\omega_1 = \omega_2 > \omega_3$). (The couplings of l_1 with $\Delta N = 2$ are of higher order in the deformation and have been neglected.)

refers to a prolate potential with the 3-axis as symmetry axis. For $\omega_{rot} = 0$, the orbits are specified in the quantum number n_3 and are $(N + 1 - n_3)$-fold degenerate. For $\omega_{rot} \approx \omega_2 - \omega_3$, a transition takes place to a coupling scheme characterized by the quantum numbers n_1 and Λ_1 ($= l_1$). The asymptotic slope of the eigenvalues in fig. 6 corresponds to $\partial\varepsilon'/\partial\hbar\omega_{rot} = -\Lambda_1$.

Orbits that are eigenstates of l_1 have a density that is axially symmetric about the 1-axis. Thus, if one lets the shape of the potential adjust self-consistently to the density distribution of the particles, the nuclear shape moves away from prolate symmetry about the 3-axis and approaches oblate symmetry about the 1-axis, through a succession of triaxial shapes. For oblate symmetry about the axis of rotation, the particles are fully aligned and yield the maximum angular momentum that can be generated by the particles in the shell. (Since the number of particles in unfilled shells is typically of order $A^{2/3}$, with each particle carrying an angular momentum of order $A^{1/3}$, this maximum value of I is of order A.) The alignment effect can thus lead to a termination of the rotational band, a phenomenon that appears to be important for band structure in the light nuclei. (All these features of particle motion in a rotating harmonic-oscillator potential are implicit in the SU_3 classification.)

The harmonic-oscillator potential represents a situation in which the rotational alignment takes place uniformly for all the orbits (corresponding to the fact that the orbits of the particles can be analysed in terms of the alignment of the individual quanta of the harmonic-oscillator motion). Large deviations from this uniformity arise from the modification in the particle motion associated with the spin-orbit coupling and the effect of the surface. An example is shown in fig. 7, which gives the eigenstates of the Hamiltonian (32) for a prolate potential with parameters appropriate to neutrons in a nucleus with $A \approx 150 \div 200$.

The Coriolis term proportional to j_1 violates the assumed axial symmetry and time reversal invariance of H_{sp}, but the Hamiltonian (32) remains invariant with respect to space reflection $\mathscr{P}$ and with respect to a rotation $\mathscr{R} = \mathscr{R}_1(\pi)$ of 180° about the axis of rotation. The eigenvalues of $\mathscr{R}$ for a single nucleon are

$$r = \exp[-i\pi j_1] = \begin{cases} -i & (j_1 = 1/2, 5/2, 9/2, \ldots)\,, \\ +i & (j_1 = 3/2, 7/2, 11/2, \ldots)\,, \end{cases} \tag{33}$$

and in the figure the states with $r = -i$ and $r = +i$ are drawn with full and dashed curves, respectively. The parity values can be found by following the orbits to $\omega_{rot} = 0$, where they are labelled by the conventional asymptotic quantum numbers $Nn_3\Lambda\Omega$; thus $\pi = (-1)^N$. (The symmetries $\mathscr{P}$ and $\mathscr{R}$ are not contingent upon the axial symmetry of the potential chosen in fig. 7, but apply to a much wider class of potentials including arbitrary ellipsoidal shapes.)

In the limit of $\omega_{rot} \to 0$, the single-particle configurations and the associated rotational band structure correspond to those considered earlier for axially and $\mathscr{R}$-symmetric nuclei. In this limit, the single-particle states are twofold degenerate, with opposite values of r. Thus, for an even-even nucleus,

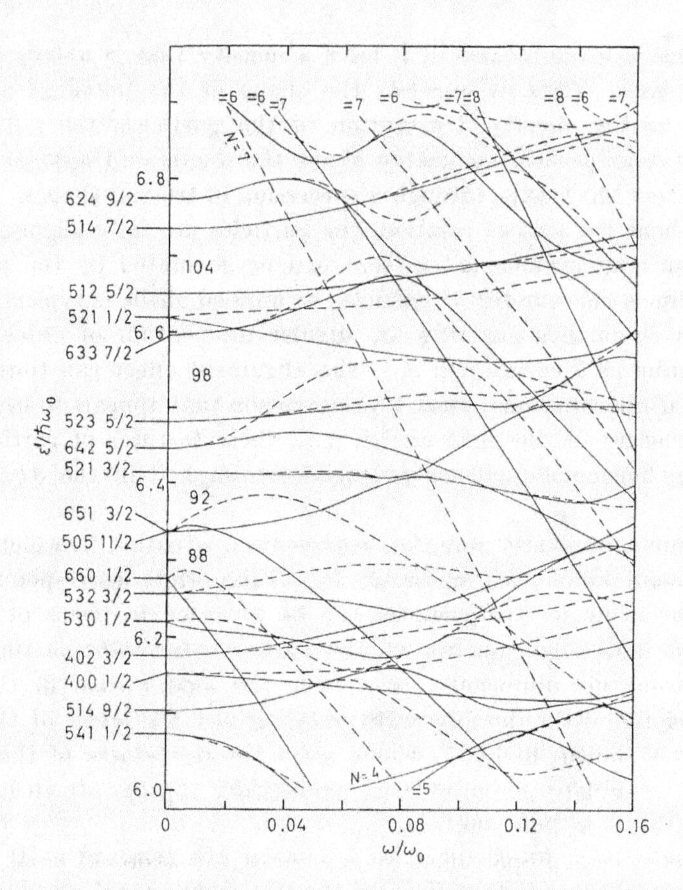

Fig. 7. – Eigenvalues of cranked single-particle Hamiltonian (32) for modified harmonic-oscillator potential. The potential is assumed to be prolate with a deformation $\delta \approx 0.25$, and the $(\boldsymbol{l})^2$ and $(\boldsymbol{l}\cdot\boldsymbol{s})$ terms have coefficients appropriate to the neutrons in the region of the rare earths. The eigenvalues ε' as well as the rotational frequency are given in units of the mean oscillator frequencies ω_0; the angular momentum I is approximately proportional to ω_{rot} and of order 65 for $\omega_{\text{rot}}/\omega_0 \approx 0.1$. The figure represents unpublished results by R. Bengtsson and S. E. Larsson to whom I am indebted for a private communication.

the ground state has $K = 0$, $r = +1$ and gives rise to a band with $I = 0, 2, 4 \ldots$. For an odd-A nucleus, the lowest configuration involves the two degenerate orbits for the last odd particle that together form a band with $I = K, K+1, K+2, \ldots$. The two orbits with specified r (that are linear combinations of the orbits with specified j_3) can be associated with the two values of the signature (8)

$$\left\{\begin{aligned} r &= \exp[-i\pi I]\,, \\ \sigma &= r^* \exp[i\pi K]\,, \end{aligned}\right. \tag{34}$$

corresponding to the fact that, in the cranking model, I is identified with the component J_1 of the total angular momentum of the particles; for the many-particle cranked state, the operation $\mathscr{R}$ equals $\exp[-i\pi J_1]$ and, for the configuration considered, has the eigenvalue r of the last orbit. Thus, the separation between the degenerate orbits that occurs for finite ω_{rot} gives the signature-dependent term in the rotational energy (see (13)). In particular, the $K=1/2$ orbits exhibit a separation that is linear in ω_{rot} and the coefficient of which yields the value of the coefficient A_1 in (13). (The operation $\mathscr{R}$ in the cranking model is associated with a rotation about the direction of $\boldsymbol{I}$, which in this context is treated as a classical vector. This operation differs from the rotations about intrinsic axes considered earlier. Thus, the quantal state (6) is not an eigenstate of $\mathscr{R}_1(\pi)=\exp[-i\pi I_1]$; in fact, in this state, all intrinsic axes perpendicular to the 3-axis are equivalent.)

The level scheme in fig. 7 gives the order of filling of the particle orbits in the rotating nucleus. While H_{sp} represents the energy of the particle, the eigenvalues of H' give the energy associated with the addition of a particle, for fixed I. In fact, if a particle is added in an orbit with nonvanishing $\langle j_1\rangle$, the collective rotational energy is reduced by the amount $\hbar\omega_{\text{rot}}\langle j_1\rangle$ (compare (30)). Each configuration gives rise to a rotational sequence with $\Delta I=2$, specified by (34) in terms of the total r value. If the shape differs from axial symmetry, additional rotational degrees of freedom may occur. These involve a precession of the 1-axis about the direction of $\boldsymbol{I}$ and, for I_1 large compared with I_2 and I_3, have the character of a vibrational mode (like the Chandler's wobbling for the Earth). The $\mathscr{R}$-symmetry condition for the spectrum is

$$r=\exp[-i\pi I_1]=\exp[i\pi(n-I)]\,, \tag{35}$$

where n is the vibrational quantum number.

A striking feature of fig. 7 is the strong rotational alignment effect that sets in for relatively small rotational frequencies, for the orbits with large j and small Ω, such as the [660 1/2] and [651 3/2] orbits that are predominantly $i_{13/2}$. These orbits are especially sensitive to the Coriolis coupling, due to the small separation between the coupled states and the large matrix elements of j_1, and the alignment is seen to occur for much smaller values of ω_{rot} than in fig. 6. (The characteristic value of ω_{rot} in fig. 6 is $\frac{1}{2}(\omega_2-\omega_3)\approx\frac{1}{2}\omega_0\delta$ and thus corresponds to $\omega_{\text{rot}}/\omega_0\approx 0.12$ in fig. 7. For orbits with large j and small Ω, the energy separations between states coupled by the Coriolis force are smaller than $\delta\hbar\omega_0$ by a factor of order j, and the matrix elements of j_1 are of order j.)

In an odd-A nucleus, in which the odd particle occupies one of the orbits with large j and small Ω, the strong rotational alignment effect gives rise, already for relatively small I values, to a transition from the coupling scheme corresponding to adiabatic rotational motion to one in which the angular momentum of the odd particle is aligned along the axis of rotation (decoupled bands [12]). These spectra (which occur especially in the transition regions between nuclei with spherical and deformed equilibrium shapes) will be discussed in [13].

For an even-A nucleus, it is important in the analysis of the corresponding rotational alignment effects for moderate values of I to take into account the pair correlations. In fact, this correlation prefers to have the particles pairwise occupying orbits that are conjugate under time reversal (or $\mathscr{R}$) and thus counteracts the rotational alignment effect that separates these orbits. (We shall return later to the competition between pairing and rotational alignment.)

Other orbits in fig. 7 are much less affected by the rotation and retain the r-doublet degeneracy up to quite large values of ω_{rot}. Thus, for example, the spectrum of ^{176}Hf shows sets of bands with only small signature effects and well-defined quantum number K [14]; these bands can be associated with quasi-particles in the neutron orbits [512 5/2], [514 7/2] and [624 9/2] that occur near the Fermi surface for neutron number 104, and corresponding proton orbits.

For large rotational frequencies, the alignment effect is seen to bring orbits from higher shells down to the Fermi surface (see, for example, the orbits in fig. 7 labelled [770 1/2] that become aligned with $j_1 = 15/2$ and $13/2$). The intersection of these orbits with more horizontal ones will imply a rather large increase in I for given ω_{rot} and a corresponding decrease of ω_{rot} for given I (backbending). For sufficiently large ω_{rot}, these orbits from the higher shells may be unbound ($\varepsilon = \langle H_{sp}\rangle = \varepsilon' + \hbar\omega_{rot}\langle j_1\rangle > 0$), when they come to the Fermi surface (see fig. 5).

A challenge for the coming years will be the interpretation of the growing evidence on yrast spectra in terms of the cranked single-particle diagrams, such as the example in fig. 7. These diagrams not only describe the single-particle configurations involved, but also provide the key to the associated rotational and vibrational degrees of freedom, and to the possible occurrence of major shell structure effects for certain values of the rotational frequency.

In the analysis of the nucleonic configurations as a function of rotational frequency, one must not only take into account that the equilibrium shape (and the pairing) varies with ω_{rot}. It must also be recognized that the nuclear potential itself, for given shape, may be a function of ω_{rot}, as already indicated by the evidence on the Coriolis coupling for small rotational frequencies. In addition to the terms in the pair potential discussed in [5], there may be new spin-dependent terms, such as those arising from the coupling of the spin of a nucleon to the orbital motion of the other nucleons (through the two-body spin-orbit force) and from the spin-spin interaction (see the discussion in ref. [11]).

6. – Rotation about symmetry axis.

As discussed above, the macroscopic centrifugal effects embodied in the liquid-drop model imply a tendency towards oblate symmetry with respect to the axis of rotation. A trend towards symmetry about the axis of rotation was also seen to be associated with the rotational alignment of the orbits of individual nucleons. This symmetry may thus play an important role in the yrast region; moreover, it provides an instructive limiting coupling scheme for the study of rotational motion.

If the angular momentum is in the direction of a symmetry axis, the angular-momentum component $\Omega = j_1$ of the individual nucleons is a constant of the motion, and the total angular momentum is the sum of the contributions of the individual nucleons:

$$I(= J_1) = \sum_{k=1}^{A} \Omega_k \,. \tag{36}$$

For rotations about the symmetry axis, the average potential and density remain stationary. Nevertheless, it is convenient in the analysis of the spectrum to consider the eigenvalues of the « cranked » Hamiltonian

$$\varepsilon' = \varepsilon - \hbar\omega_{\text{rot}}\Omega \tag{37}$$

which yield a diagram of straight lines in the plot of ε' as a function of ω_{rot} (see fig. 8). Thus, for any value of ω_{rot}, the set of lowest ε' values represents an yrast configuration; in fact, any other configuration with the same I will have a larger value of $\sum \varepsilon'$ for the value of ω_{rot} considered and, hence, also a greater value of $\sum \varepsilon$. Such an yrast configuration can be represented as a « vacuum » state over a certain region of ω_{rot}; when a crossing occurs at the Fermi surface between an occupied and an empty orbit, a new vacuum state is formed with an increase in I corresponding to the difference between the Ω values of the crossing levels. The yrast levels with intermediate values of I are obtained as particle-hole (or vibrational, or rotational) excitations.

The states on the yrast line are connected by changes in the single-particle configuration. Hence, the possibility arises of high-spin yrast states with lifetimes much greater than those characteristic of collective rotational transitions. The observation of such « yrast traps » and the study of their decay would provide favourable new opportunities for yrast spectroscopy. (For an analysis of yrast states in specific nuclei expected to have oblate shape in regions of I, see, for example, [15].)

The average slope of the yrast line $\mathscr{E}(I)$ can be obtained by a statistical analysis based on an average single-particle level density $g(\Omega)$ at the Fermi

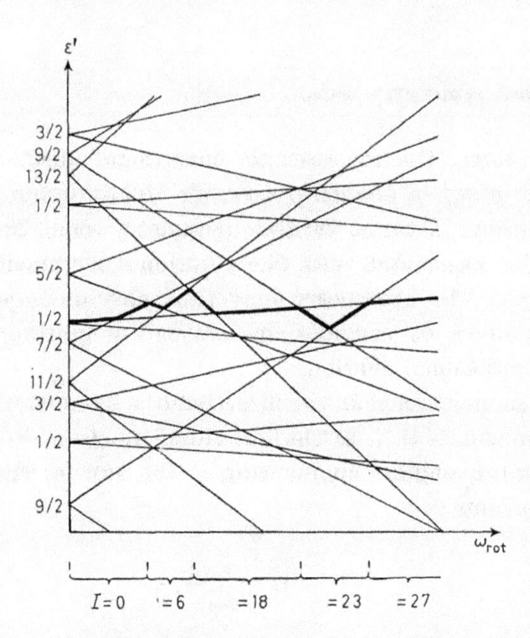

Fig. 8. – Eigenvalues for single-particle Hamiltonian cranked about a symmetry axis. The spectrum for $\omega_{rot}=0$, labelled by the quantum number Ω, is schematic and for illustrative purposes only.

surface. For each ω_{rot}, the occupied orbits are those with $\varepsilon' \leqslant \lambda'$, corresponding to

$$\varepsilon(\Omega) \leqslant \lambda(\Omega) = \lambda' + \hbar\omega_{rot}\,\Omega \,. \tag{38}$$

Since $g(-\Omega) = g(\Omega)$, the total angular momentum is given by

$$I = \sum_{occ} \Omega = \sum_{\Omega} \Omega(\lambda(\Omega) - \lambda')\, g(\Omega) = \hbar\omega_{rot} \sum_{\Omega} \Omega^2 g(\Omega) \,, \tag{39}$$

while the energy is

$$\left\{\begin{aligned} \mathscr{E} &= \sum_{occ} \varepsilon = \mathscr{E}(\omega_{rot} = 0) + \mathscr{E}_{rot}, \\ \mathscr{E}_{rot} &= \sum_{\Omega} \frac{1}{2} (\lambda(\Omega) - \lambda')^2 g(\Omega) = \frac{1}{2} (\hbar\omega_{rot})^2 \sum_{\Omega} \Omega^2 g(\Omega) = \frac{\hbar^2}{2\mathscr{I}} I^2 \end{aligned}\right. \tag{40}$$

with

$$\mathscr{I} = \hbar^2 \sum_{\Omega} \Omega^2 g(\Omega) = \frac{\hbar I}{\omega_{rot}} \,. \tag{41}$$

Since the average value of $|\Omega|$ at the Fermi surface increases as $A^{1/3}$, while the total level density $\sum_{\Omega} g(\Omega)$ is proportional to A, the effective moment of inertia (41) is seen to be of order $A^{5/3}$, in units of Mr_0^2, where r_0 is the radius parameter ($R = r_0 A^{1/3}$). The moment of inertia is thus of the order of the rigid-body value. In fact, an evaluation of the expression (41) on the basis of the Fermi-gas model yields precisely the rigid-body value. This result reflects the fact that the distribution of particle orbits is such as to yield an isotropic velocity distribution and hence no net flow, in the rotating co-ordinate system.

Thus, for rotation about a symmetry axis, the average energy as a function of angular momentum is the same as for collective rotation. The two modes of rotation, however, are characterized by radically different yrast spectra.

7. – Pairing in rotating potentials. Phase transition.

As discussed above, the pair correlations lead to a decrease in the rotational moment of inertia and, hence, to an increase in the rotational energy for given I. Thus, for sufficiently large rotational frequencies, the gain in energy associated with the pair correlation is upset by the increased rotational energy, and one expects a phase transition to normal nuclear matter [16]. The study of the manner in which this phase transition takes place is a topic of great current interest, as reflected in the lively publication activity in the nuclear-physics journals.

With pairing included, single-particle motion in a rotating potential can be described by a Hamiltonian of the form

$$H' = H_{sp} + V_{pair} - \lambda\mathcal{N} - \hbar\omega_{rot} J_1 \tag{42}$$

with

$$H_{sp} - \lambda\mathcal{N} = \sum_{\nu} (\varepsilon(\nu) - \lambda)\, a^\dagger(\nu)\, a(\nu)\,, \tag{43}$$

$$V_{pair} = -\Delta \sum_{\nu>0} (a^\dagger(\bar\nu)\, a^\dagger(\nu) + a(\nu)\, a(\bar\nu))\,, \tag{44}$$

$$J_1 = \sum_{\nu_1\nu_2} \langle \nu_2 | j_1 | \nu_1 \rangle\, a^\dagger(\nu_2)\, a(\nu_1)\,, \tag{45}$$

where ν labels the eigenstates of H_{sp} and $\bar\nu$ is the time reverse of ν. The number operator is denoted by $\mathcal{N}$. The pair potential (44) includes only the monopole term that creates and annihilates pairs of states conjugate under time reversal; additional terms in the pair potential may be present in the rotating nucleus (see [5]). The strength Δ of the pair potential, as well as the shape of the nucleus, is a function of ω_{rot} characterizing the equilibrium for given rotational frequency.

The Hamiltonian (39) is a bilinear form in the particle creation and annihilation operators $a^\dagger$, a and can be brought to diagonal form by a linear transformation to quasi-particle operators

$$\alpha^\dagger(i) = \sum_\nu \left(u_{i\nu} a^\dagger(\nu) + v_{i\nu} a(\bar{\nu})\right) , \tag{46}$$

$$H' = \text{const} + \sum_i E'_i \alpha^\dagger(i)\alpha(i) . \tag{47}$$

The transformation (46) is a generalization of the more familiar one which applies to time-reversal–invariant potentials. Essential new features are that the quasi-particle states no longer have the twofold degeneracy (but retain the quantum numbers π and r, if the symmetries $\mathscr{P}$ and $\mathscr{R}$ apply to H_{sp}, since the pair potential (44) is invariant under these operations), and that the quasi-particle energies E' can be smaller than Δ (in analogy to the situation in gap-less superconductors [17]).

We can see the new features most easily for a nucleus rotating about the symmetry axis. In this case, the eigenstates ν of H_{sp} are also eigenstates of j_1 (with eigenvalue Ω). The quasi-particle transformation is now the usual one, as for $\mathscr{T}$-invariant potentials,

$$\left\{ \begin{aligned} \alpha^\dagger(\nu) &= u_\nu a^\dagger(\nu) + v_\nu a(\bar{\nu}) , \\ a^\dagger(\nu) &= u_\nu \alpha^\dagger(\nu) - v_\nu \alpha(\bar{\nu}) , \end{aligned} \right. \tag{48}$$

which leaves the operator J_1 diagonal

$$J_1 = \sum_\nu \Omega a^\dagger(\nu) a(\nu) = \sum_\nu \Omega \alpha^\dagger(\nu)\alpha(\nu) , \tag{49}$$

and the quasi-particle energies are

$$E'(\nu) = \left((\varepsilon(\nu) - \lambda)^2 + \Delta^2\right)^{\frac{1}{2}} - \hbar\omega_{rot}\Omega . \tag{50}$$

The quasi-particle spectrum (50) is illustrated schematically in fig. 9. For an even number of particles, the quasi-particle vacuum ($v = 0$) is the lowest state for rotational frequencies that are smaller than the value for which the sum of the two lowest quasi-particle energies vanishes. For larger ω_{rot}, this two–quasi-particle state becomes the lowest (the yrast state), until the next pair of quasi-particles has zero energy, after which the four–quasi-particle state moves to the yrast line, etc. It is seen that the characteristic frequency ω_1 for the first of these crossings is of order

$$\omega_1 \sim \frac{\Delta}{\Omega_{max}} , \tag{51}$$

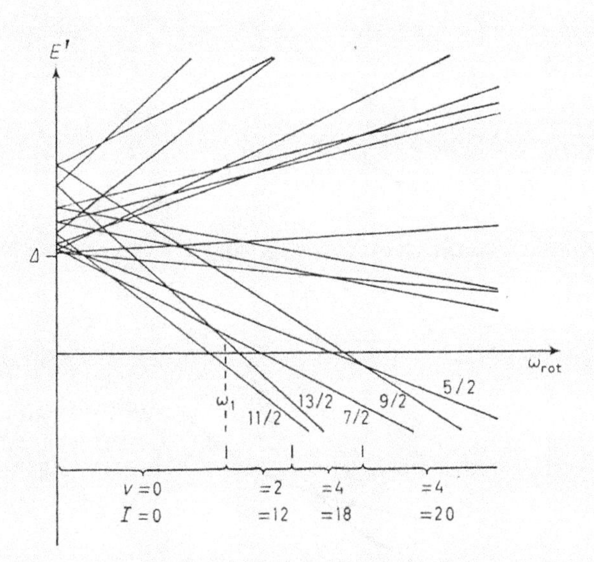

Fig. 9. – Quasi-particle energies E' corresponding to the schematic single-particle spectrum in fig. 8.

where $\Omega_{\max}$ $(= j_{\max})$ is the largest single-particle angular momentum near the Fermi surface.

The possibility of negative-energy quasi-particle excitations appears as a general feature of the pairing in rotating potentials. In fact, the rotational alignment effect implies that quasi-particles carry a nonvanishing component $\langle j_1 \rangle$ of angular momentum; thus, the excitation of a quasi-particle, for fixed I, is associated with a decrease in the collective rotational energy, corresponding to the last term in (47). When the sum of two quasi-particle energies vanishes, one expects a band crossing on the yrast line. For example, in an even-even nucleus, a $v = 2$ band with a large value of $\langle j_1 \rangle + \langle j_2 \rangle$ may cross the $v = 0$ band (This interpretation of the observed backbending effect is equivalent to that given in [12].)

The intersection of two bands is schematically illustrated in fig. 10. The upper part of the figure is a plot of I *vs.* ω_{rot}; in the cranking diagram (as in fig. 9), the two bands intersect $(\mathscr{E}'_1 = \mathscr{E}'_2)$ at some frequency $(\omega_{\text{rot}})_0$, and this crossing is associated with an increase of I from I_1 to I_2. In the $\mathscr{E}$-I diagram (where $\mathscr{E} = \mathscr{E}' + \hbar\omega_{\text{rot}} I$), the band crossing takes place at the intermediate angular momentum $I_0 \approx \frac{1}{2}(I_1 + I_2)$, as illustrated in the lower part of fig. 10. In a plot of I or of the moment of inertia $\mathscr{I} = \hbar I/\omega_{\text{rot}}$ against ω_{rot}, the yrast line will thus backbend at $I = I_0$.

The sharp crossing of levels as in fig. 9 is a special feature of the symmetric potential, for which Ω is a constant of the motion for the quasi-particles. In the absence of this symmetry, there will be interactions between the crossing bands, although a weak one if the interacting quasi-particle states have very different values of $\langle j_1 \rangle$. In the cranked potential, a weak interaction will lead

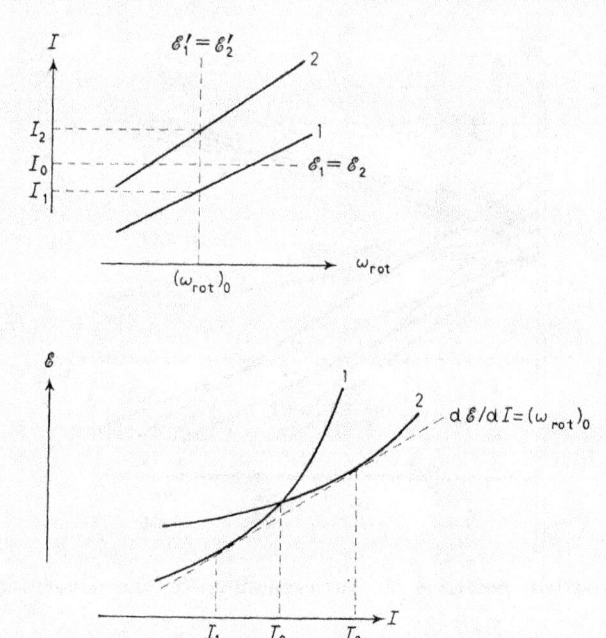

Fig. 10. – Intersection of bands in terms of the variables ω_{rot}, I, $\mathscr{E}'$ and $\mathscr{E}$.

to a rather abrupt transition of the yrast line from state 1 to state 2 in the region of $(\omega_{\text{rot}})_0$, which would correspond to an abrupt transition in the $\mathscr{E}$-I diagram from I_1 to I_2 along the straight dashed line. Actually, the intersection of the bands takes place in the region of I_0, where the bands come close together, for the same value of I. A proper treatment of this interaction thus goes beyond the cranking model in its most simple form, since the two interacting bands are associated with two different rotational frequencies. (For a discussion of this problem, see [18].)

The appearance of states along the yrast line with an increasing number of quasi-particles leads at the same time to a systematic decrease in the pair correlations, since the states occupied by the quasi-particles are no longer available to the pair correlations (blocking effect). Both these effects appear as essential features of the phase transition.

The variation of Δ with ω_{rot} can be explicitly evaluated for rotations about the symmetry axis (B. MOTTELSON: private communication). In the absence of rotation, the self-consistency condition for Δ is given by

$$G\sum_{\nu>0}\frac{1}{2E_\nu}=1 \tag{52}$$

$$=\frac{1}{4}gG\int_{-S/2}^{S/2}\frac{d\varepsilon}{(\varepsilon^2+\Delta^2)^{\frac{1}{2}}}$$

$$=\frac{1}{2}gG\log\frac{S}{2\Delta}\,,$$

where G is the pairing coupling constant. In the second line of (52), we have assumed a uniform level spectrum with total density g ($= 2/d$, where d is the distance between the doubly degenerate single-particle states) and have assumed the pairing to act with constant matrix element within an energy interval $S/2$ on either side of the Fermi level.

In the rotating system, the single-particle levels with negative E' are blocked and are to be removed from the sum in (52). Thus, only levels with

$$\hbar\omega_{\text{rot}}|\Omega| < ((\varepsilon - \lambda)^2 + \Delta^2)^{\frac{1}{2}} \tag{53}$$

are to be included (see (50)), and the evaluation of the resulting integral leads to an equation for $\Delta(\omega_{\text{rot}})$. In particular, the critical frequency ω_{crit} for which $\Delta \to 0$ is found by replacing the total level density g by

$$\sum_{|\Omega|\hbar\omega_{\text{crit}} < |\varepsilon-\lambda|} g(\Omega) = g f\left(\frac{|\varepsilon - \lambda|}{\hbar\omega_{\text{crit}}\Omega_{\text{max}}}\right) \tag{54}$$

for $|\varepsilon - \lambda| < \hbar\omega_{\text{crit}}\Omega_{\text{max}}$. The dimensionless function $f(x)$, which depends on the distribution of levels over Ω, increases from $f(x = 0) = 0$ to $f(x = 1) = 1$. The evaluation of the integral in (52) with $\Delta = 0$ now yields

$$\left\{\begin{aligned} &\int_{-S/2}^{S/2} = 2\left(\int_0^{\varepsilon_1}\frac{d\varepsilon}{\varepsilon} f\left(\frac{\varepsilon}{\varepsilon_1}\right) + \int_{\varepsilon_1}^{S/2}\frac{d\varepsilon}{\varepsilon}\right) = 2\left(k + \log\frac{S}{2\varepsilon_1}\right), \\ &\varepsilon_1 = \hbar\omega_{\text{crit}}\Omega_{\text{max}}, \end{aligned}\right. \tag{55}$$

where k is a number of order unity. Since the integral (55) is equal to its value for $\omega_{\text{rot}} = 0$, which is given by (52), we obtain

$$\hbar\omega_{\text{crit}} = \frac{\Delta(\omega_{\text{rot}} = 0)}{\Omega_{\text{max}}}\exp[k], \tag{56}$$

which shows that the critical frequency is of order $\Delta/\Omega_{\text{max}}$. This is of the same order of magnitude as the value (51) of ω_{rot}, for which negative quasi-particle energies appear. (For rotations about the symmetry axis, the rotation-induced terms in the pair potential (discussed in [5]) are not present, corresponding to the fact that the condensate of pairs does not participate in the rotation. For the symmetric potential, the angular momentum is entirely carried by the quasi-particles.)

Thus, it appears that $\Delta/\Omega_{\text{max}}$ is a basic quantity characterizing the scale of rotational frequencies at which the phase transition occurs. However, the structure of the transition depends in an important way on the geometry of the system and appears to involve a variety of intriguing phenomena.

REFERENCES

[1] A. Bohr and B. R. Mottelson: *Nuclear Structure*, Vol. **2** (Reading, Mass., 1975).
[2] D. R. Bés and R. A. Broglia: this volume, p. 55.
[3] C. W. Reich and J. E. Cline: *Nucl. Phys.*, **159** A, 181 (1970).
[4] C. J. Gallagher, O. B. Nielsen and A. W. Sunyar: *Phys. Lett.*, **16**, 298 (1965); C. Günther and D. R. Parsignault: *Phys. Rev.*, **153**, 1297 (1967); J. M. Domingos, G. D. Symons and A. C. Douglas: *Nucl. Phys.*, **180** A, 600 (1972).
[5] I. Hamamoto: this volume, p. 234 and 252.
[6] B. Mottelson: this volume, p. 31.
[7] D. J. Thouless and J. G. Valatin: *Nucl. Phys.*, **31**, 211 (1962).
[8] S. Cohen, F. Plasil and W. J. Swiatecki: *Ann. of Phys.*, **82**, 557 (1974).
[9] H. C. Britt, B. H. Erskilla, R. H. Stokes, H. H. Gutbrod, F. Plasil, R. L. Ferguson and M. Blann: *Phys. Rev. C*, **13**, 1483 (1976); H. Gauvin, D. Guerrau, Y. Le Beyec, M. Lefort, F. Plasil and X. Tarrago: *Phys. Lett.*, **58** B, 163 (1975).
[10] S. Åberg: this volume, p. 473.
[11] K. Neergård, V. V. Pashkevich and S. Frauendorf: *Nucl. Phys.*, **262** A, 61 (1976).
[12] F. S. Stephens: *Rev. Mod. Phys.*, **47**, 43 (1975).
[13] F. S. Stephens: this volume, p. 172.
[14] T. L. Khoo, F. M. Bernthal, R. H. G. Robertson and R. A. Warner: *Phys. Rev. Lett.*, **37**, 823 (1976).
[15] J. Dudek: this volume, p. 465 and 469.
[16] B. R. Mottelson and J. G. Valatin: *Phys. Rev. Lett.*, **5**, 511 (1960).
[17] A. Goswami, L. Lin and G. L. Struble: *Phys. Lett.*, **25** B, 451 (1967).
[18] I. Hamamoto: *Nucl. Phys.*, **263** A, 315 (1976).

Aage Bohr in the garden of his home in Copenhagen on the occasion of the celebration of his 50th birthday in 1972. Courtesy of Niels Bohr Archive, Copenhagen.

Tale af Professor Aage Bohr
den 3. marts 1971

For os, der er knyttet til dette Institut, har denne dag en meget festlig karakter, og jeg vil gerne byde hjertelig velkommen til alle vore gæster, som fra nær og fjern er kommet til stede for at være med til at fejre jubilaren.

Som de fleste vil vide, har vi ikke indskrænket os til at fejre jubilæet på denne dag, der markerer de halvtreds år fra Institutets officielle indvielse, men har følt, at anledningen berettigede til et helt jubilæumsår.

Vi havde den særlige opmuntring at kunne indlede jubilæumsåret med indvielsen af den nye tandemaccelerator. Den lever tilfulde op til forventningerne og har givet jubilæumsåret - også forskningsmæssigt - en ekstra kulør. Ved en række jubilæumsforedrag og symposier, sigter vi på de almene problemstillinger i kvantefysikken og søger desuden at udvide vor horisont i retninger, hvor atomfysikken har kontaktflader til andre fag, som astronomi, geologi og biologi, såvel som filosofi, psykologi og arkæologi.

- 2 -

Der er idag særlig grund til at se tilbage på Instituttets oprettelse og dets indvielse. Vi har forskelligt materiale fra selve indvielsen, dels manuskriptet til min fars tale, dels referaterne i dagspressen og endelig er der jo nogle få iblandt os, der kan fortælle om begivenheden. Foruden min mor er det jo fru Schultz, der dengang allerede i flere år havde indtaget sin betroede post.

Lærestolen i teoretisk fysik blev oprettet i 1916. Der var dengang intet institut knyttet dertil, men som det hedder i indvielsestalen "Trangen dertil meldte sig straks", idet "man ved de teoretiske Undersøgelser stadig står overfor Spørgsmål om, hvorvidt de benyttede Antagelser er i Overensstemmelse med Virkeligheden,og man under Arbejdets Fremadskriden stadig er henvist til at lade Naturen selv afgøre sådanne Tvivlsspørgsmål ved Anstillelsen af Eksperimentalundersøgelser".

Det bidrog væsentligt til sagens fremme, at der fra privat side blev vist offervillig interesse for Institutets oprettelse. Grunden, hvorpå vi befinder os, blev jo skænket Universitetet ved midler, der indsamledes fra privat side, og da Institutet skulle indrettes,trådte Carlsbergfondet til. Vi hører også i indvielsestalen at "En Tid lang så det alligevel ud til, at det skulle blive helt galt for os på grund af Valutaforholdene, idet de i England bestilte Apparaters Pris steg i samme Forhold, som den danske Krone faldt". Men også denne sidste hindring blev bragt af vejen gennem ædelmodig bistand, denne gang fra et bankierfirma.

Det nyoprettede Institut blev mødt med stor velvilje fra alle sider. Pressen viede begivenheden betydelig interesse

- 3 -

og gav ret fyldige referater af indvielsen; man hører således om den tale, som blev holdt af Universitetets daværende rektor, Otto Jespersen. "Det er glædeligt", skal han have sagt ifølge Politiken, "at en sådan Institution kan rejses i denne for Materialisme så udskregne Tidsalder". Rektor udtalte videre sin tillid til Institutets leder og tilføjede "De har forstået at samle både inden- og udenlandske Forskere omkring Dem og har således på den smukkeste Måde genoptaget det internationale Samarbejde, der afbrødes af Verdenskrigen. Det er i Glædens og Forventningens Tegn, at jeg erklærer Institutet for åbent". I Socialdemokraten under overskriften "Dansk Videnskab åbner sit ny Tempel" kunne man læse en reportage, der var frisk og lødig, men ikke særlig ærbødig i sin omtale af de indbudte honoratiores, deriblandt medlemmer af regeringen. Man bemærker også følgende passus: "Man skimter allerede langt ude en ny verdensomvæltende Drivkraft, der vil revolutionere Samfundet teknisk, som Socialismen vil gøre det økonomisk".

I fars indvielsestale hører man allerede om planer og ønsker om nyt apparatur i "en ikke altfor fjern fremtid". Det varede jo også kun et par år, før der blev stærkt behov for udvidelser af bygningerne, hvortil midlerne blev skænket af "Rockefellerfondet", vist et af dette fonds første bevillinger.

Samspillet mellem teori og eksperiment, der var et hovedmotiv for Institutets oprettelse, har hele tiden været noget meget centralt, og det har været af afgørende betydning, at Institutet har kunnet finde støtte til udbygning af forsøgsapparaturet i takt med udviklingen.

- 4 -

Da atomkernefysikken kom i forgrunden, kom det til at dreje sig om store installationer, der bekostedes først af Carlsberg- og Thrigefondet, og senere ved store bevillinger fra det offentlige. I disse henseender og på så mange andre måder, er Institutet igennem alle årene blevet mødt med stor velvilje og tillid fra så mange kredse i det danske samfund. Vi føler en dyb taknemmelighed herfor og håber, at vi også i fremtiden vil kunne gøre os fortjent hertil. Måske vil nogle have fået det indtryk, at Institutet altid straks har fået, hvad det ønskede, men så let er det ikke gået, hvad de bugnende arkiver med Institutets ansøgninger vil kunne bevidne.

Som det vil være bekendt, arbejder vi i disse år med store planer om et nyt fremstød på atomkernefysikkens område og håber denne gang at forene ressourcer, både menneskelige og materielle, fra alle de nordiske lande. Vi ser store og meget spændende perspektiver i disse planer og glæder os over den brede støtte, de allerede har opnået, selvom vi naturligvis også er beredte på, at et så stort foretagende vil kunne støde på mange fødselsvanskeligheder.

Atomfysikkens udvikling i de forløbne halvtreds år er kendetegnet ikke blot ved den dybtgående udforskning af stoffets indre struktur, men tillige ved stadig større bredde og stadig nye forbindelser til andre forskningsfelter. Det store gennembrud i forståelsen af atomets opbygning skabte hurtigt et nyt grundlag for tydningen af grundstoffernes periodiske system og for analysen af de kemiske bindinger, og udviklingen førte videre til forståelsen af centrale biologiske processer på atomar basis. Studiet af de kernefysiske fænomener

skulle

allerede på et tidligt trin give helt nyt værktøj til geologien og biologien og åbne mulighed for forståelsen af stjernernes egenskaber.

Nogle af disse konsekvenser af opdagelsesrejsen i atomets verden var allerede kendte eller kunne anes for halvtreds år siden, men bestandig er man stødt på fænomener af helt uventet karakter, der har givet forskningen nye retninger.

Til de uforudsete konsekvenser af atomfysikkens udvikling hører jo også den store forøgelse af menneskets herredømme over naturkræfterne. Naturvidenskaben og dens repræsentanter blev herigennem på ny måde konfronteret med samfundsmæssige problemstillinger, og udviklingen i de sidste fem og tyve år har givet yderligere aktualitet til hele problemkredsen omkring videnskabens stilling i samfundet. I denne sammenhæng er videnskabens internationale karakter et væsentligt punkt. Det ligger jo i videnskabens natur, at den ingen grænser kender og bygger på internationalt samarbejde, men det er en opgave for fremtiden at drage de fulde konsekvenser af dette forhold, både hvad angår det videnskabelige arbejde og dets samfundsmæssige aspekter. Nødvendigheden af åbenhed mellem nationerne, som min far pegede på, synes at tegne sig stadig mere klart. For Institutet med dets placering i det internationale videnskabelige samarbejde er det en naturlig opgave at bidrage til en udvikling i denne retning gennem en videre udbygning af dette samarbejde.

I 1921 var det klart, at beskrivelsen af de atomare

- 6 -

fænomener førte ud over de klassiske fysiske årsagssammenhænge. Klarlæggelsen af disse forhold skulle i Institutets første tiår føre til en ny belysning af menneskets stilling som iagttager af den natur, hvoraf det selv er del. Denne nye erkendelse førte også ind på problemer, som i det traditionelle universitet hører hjemme i de humanistiske fakulteter, og der ligger her kim, som venter på grobund til videre udfoldelse. Et af symposierne i jubilæumsåret tager netop sigte på en belysning af problemer på disse grænseområder.

Også i spørgsmålet om forskningens sammenknytning med undervisningen lå der i indvielsestalen for halvtreds år siden en klart formuleret holdning. Denne sammenknytning blev fremhævet som "et Moment af største Betydning som Forvarsel for Institutets lykkelige Trivsel", idet "den Opgave at skulle føre et stadigt fornyet Antal yngre Mennesker ind i Videnskabens Resultater og Metoder bidrager i højeste Grad til stadig at tage Spørgsmål op til Drøftelse fra nye Sider, og ikke mindst føres der jo igennem de unge Menneskers egen Indsats stadig nyt Blod og nye Tanker ind i Arbejdet".

Denne side af Institutets arbejde har da også været afgørende for den stadige fornyelse af hele virksomheden, og ofte er meget værdifuldt initiativ kommet fra de yngste medarbejdere og fra de studerende. Det gælder også i disse år, blandt andet i forbindelse med de aktuelle opgaver Institutet er stillet overfor ved tilrettelæggelsen af fysikundervisningen under hensyntagen til de mange nye aspekter af naturvidenskabens stilling i samfundet.

- 7 -

Min faders arbejdsform involverede jo altid et kollektiv, hvad enten det drejede sig om udarbejdelsen af et brev, opførelsen af et institut eller om selve det videnskabelige arbejde. Vi hører i indvielsestalen om den store kreds, der var inddraget i planlægningen og indretningen af Institutet. Hvad angår det videnskabelige udstyr var det jo særlig daværende docent H.M. Hansen, der så beredvillig hjalp ved "Gennemtænkningen af selv de mindste Detailler", men også James Franck's store erfaringer blev inddraget i planlægningen. Vi hører også om arkitekterne, hvis "Opfindsomhed og Taalmodighed Gang på Gang er blevet sat på den stærkeste Prøve", og om den højt værdsatte indsats fra alle håndværkernes side.

Institutet med alle dets medarbejdere blev da også fra begyndelsen en sammenhængende organisme med stærkt helhedspræg. Som andre organismer har det haft stadig trang til at vokse, og det har været opgaven så godt som muligt at bevare helhedspræget. Vi har jo endvidere i de senere år arbejdet med nye organisationsformer for at nå frem til den bedst mulige udnyttelse af det samlede initiativ i hele den store medarbejderkreds. Også på dette felt søger vi at kombinere teori med eksperiment for ligesom at spørge den menneskelige natur til råds.

Idag går vore tanker i særlig grad til de mange tidligere medarbejdere, der igennem længere eller kortere perioder har virket ved Institutet og ydet så værdifulde bidrag til arbejdet. Vi er glade for her at se så mange medarbejdere, der nu virker ved andre institutioner herhjemme. Dette slægtsskab har bidraget til at styrke samhøringheden inden for hele

- 8 -

kredsen af danske fysikere, som vi gennem årene har kunnet glæde os over.

Den første medarbejder, der kom hertil fra de andre nordiske lande, var Oscar Klein, der allerede var med i arbejdet ,før Institutet var bygget,og som jo senere i en lang årrække var fars nærmeste og uvurderlige medarbejder. Vi er meget glade for, at både Oscar og Gerda Klein har kunnet være med os idag, og vi hilser også hjertelig velkommen til vore andre nordiske medarbejdere, som er kommet hertil idag. Det nordiske islæt i Institutets liv har jo spillet en meget fremtrædende rolle gennem alle årene, og det var også baggrunden for oprettelsen af NORDITA, vor søsterorganisation, som vi lever med i en symbiose, der er af så stor betydning for hele miljøet omkring Institutet.

Vore tanker går også videre til den store kreds af medarbejdere, ialt op mod tusinde, fra alle dele af verden, som har deltaget i Institutets virksomhed og tilført det så stor inspiration. Vi havde gerne benyttet lejligheden til at samle så mange som muligt til drøftelse af aktuelle problemer, men pladsforholdene på Institutet ville ikke strække til, og et møde med så mange deltagere ville næppe heller passe til vor vante diskussionsform. Vi har derfor måttet nøjes med at sende dem alle den lille hilsen, som også er omdelt idag.

Vi kan imidlertid glæde os over den stadig levende kontakt inden for hele denne kreds, og vi fornemmer, at alle føler sig som medlemmer af een stor familie. Institutets nære

- 9 -

bånd til det internationale samfund af fysikere og de mange friske kræfter, som hvert år søger hertil fra alle dele af verden, er jo blandt de aktiver, hvortil vi knytter de største forhåbninger til fremtiden.

Far plejede at sige om begrebet lykke, at det betegner en situation, hvor det går èn bedre,end man egentlig havde fortjent, og jeg tror han selv nærede den følelse overfor sit virke. Når vi som idag ser på de halvtreds år, må vi sige, at de står som noget af et eventyr, et af dem, hvori forventningerne ofte bliver overtruffet. Og ser vi fremad, er der nok af opgaver at tage fat på og spændende perspektiver, der tegner sig forude, og mange har vi at støtte os til. Lad os håbe, at man om halvtreds år vil kunne sige, at vi havde lykken med os.

1/3-71
AB/LBM

Speech by Professor Aage Bohr (3rd March 1971)

Translated by Rob Sunderland (Niels Bohr Archive)

For those of us who are connected to this institute, today has a celebratory air, and I would like to give a hearty welcome to all of our guests, who have come from near and far to be here to celebrate our jubilee.

As most will know, we have not limited ourselves to merely celebrating the 50th anniversary of the Institute's opening day, but have felt ourselves deserving of a whole jubilee year.

We were especially encouraged to commence our jubilee year with the opening of the new tandem accelerator. It lives up to our fullest expectations and has given this jubilee year — also in terms of research — extra colour. With a series of jubilee lectures and symposia, we address the everyday problems of quantum physics and try to expand our horizons to the area which atomic physics shares with other subjects, such as, astronomy, geology and biology as well as philosophy, psychology and archaeology.

Today, there is a special reason to look back at the Institute's foundation and opening day. We have a variety of material from the opening day; the manuscript of my father's speech, cuttings from the press and finally, there are some among us today who can tell us about the events that day. Besides my mother, there is Mrs. Schultz, who already then had been at her trusted post for several years.

The chair of Theoretical Physics was established in 1916. At the time, there was no institute connected to it, but as it says in the opening speech; "The need for one soon became apparent", because, "the theoretical studies still pose questions as to whether the assumptions employed are concurrent with reality and during the progression of the work one is still compelled to let nature herself decide these conundrums by recourse to experimental investigation."

It was of considerable benefit to the cause that private donators showed such generosity towards and interest in the Institute's foundation. The grounds on which we now find ourselves were donated to the University with funds from private donations and when the institute was furnished, the Carlsberg Foundation stepped in. We also hear from the opening speech that, "for a long time, it looked like that the rates of exchange would ruin everything for us , as the apparatus we had ordered in England's price rose as the value of the Danish Krone fell". This last obstacle was removed through the noble support of a firm of bankers. The newly established institute was met with great goodwill from everywhere. The press showed particular interest and reported fully on the opening day; we hear of the speech given by the University's rector at the time, Otto Jespersen. "It is encouraging", he was reported as having said "that such an institution can be built in what for materialism is such a meagre age." The rector continued by expressing his confidence in the Institute's leader and added; "You have managed to gather both native and foreign researchers around you and have, in the most beautiful way, resumed that international collaboration which was interrupted by the World War. It is in a spirit of joy and expectation that I declare this institute open". In the "Social Democrat" under the headline, "Danish Science opens its new Temple", was an article that was fresh and incisive, but not especially respectful in its description of the invited dignitaries, amongst whom were members of the government. The following passage in particular grabs our attention: "On can already spy on the horizon a new world changing power that will revolutionise society technically as socialism will do economically".

In my father's opening speech, one hears also of his plans and wishes for new apparatus for the "not too distant future". It was only to be a couple of years before there was a pressing need for expansion of buildings to which the Rockefeller Foundation contributed with what was one of its first bequests.

The interplay between theory and experiment, which was the primary motivation for the Institute's foundation, has always been at its core and this has been of primary importance for the Institute in its search for support for building projects and the construction of experimental apparatus dictated by progress.

As atomic physics came to the fore, large installations became the order of the day. These were payed for by the Carlsberg and Thrige foundations and later by large bequests from the state. In these cases and in many others, the Institute has been treated through the years with a large portion of goodwill and trust from so many circles in Danish society. We feel a deep gratitude for this and hope that we will also be deserving of this in the future. It maybe that some may feel that the Institute has always promptly got everything it wished for, however, it has not been that easy as the fulsome archives of the Institute's applications for funds can attest.

As may be known, these days we are working on great plans for a new assault on the field of atomic physics and hope to unify the resources, both human and material, of all the Nordic countries. We see great perspective for these plans and are encouraged by the broad support they have already received though we must be aware that such a great endeavour will experience many birth pangs.

The development of atomic physics over the last fifty years has not only been characterised by thorough research of the inner structure of matter, but also by ever greater and newer collaborations with other disciplines. The great breakthrough in the understanding of the structure of the atom has rapidly shaped a new foundation for the detection of the periodic system of elements and for analysis of the chemical bonds, a development which has led to an understanding of fundamental biological processes at an atomic level. The study of nuclear physical phenomena gave a whole new tool to geology and biology and opened the possibility to understand the structure of the stars.

Some the consequences of the voyage of discovery into the atom were already known, or could be foreseen fifty years ago, but underway we have encountered phenomena of unexpected kinds which have set our research on new courses.

As for the unexpected consequences of the development of atomic physics, they include some of the great expressions of man's dominion over nature. Natural science and its representatives were, because of these, confronted with social questions and the development over the last twenty five years has given a greater urgency to the question of science's place in society. In this context, the international character of science has been an important factor. It is the nature of science that it does not recognise frontiers and builds upon international cooperation, but it is a task for the future to address the fullest consequences of this relationship pertaining to both scientific work and social aspects. The necessity of openness between nations, which my father pointed out, seems ever more obvious. For the Institute, with its position in international scientific endeavour, it is natural to contribute to such developments by enhancing such collaboration.

In 1921 it was clear that the description of atomic phenomena went beyond the classical physical causality. The clarification of these relations were, during the Institute's first decade, to lead to a new illumination of mankind's position as an observer of that nature of which he himself is part. This new acknowledgement also led to problems, which in a traditional university belong to the humanities faculties, and here lie the seeds waiting for a fertile soil in which to grow. One of our symposia in this jubilee year takes aim at addressing these problems at the disciplinary frontiers.

Referring to the question of the relationship between research and teaching, a clear opinion lay in the opening speech fifty years ago. The relationship was promoted as "a matter of extreme importance for predicting the Institute's happy well-being", in that "this task, to lead an ever renewed group of young people into the results and methodology of science, contributes at the highest level to ensure that questions are examined from new angles and not least, that through the participation of young people, new blood and thinking is brought to the endeavour." This side of the Institute's work has also been decisive for the regular renewal of the whole endeavour and it has often been the case that extremely valuable initiatives have come from the youngest colleagues and students. This is particularly true these days, amongst other things due to the questions addressed by the Institute which have meant an adaption of our teaching to include the many aspects of science's place in society.

My Father's work methods always involved a collective, whether in the composition of a letter, the foundation of an institute or the scientific work itself. In the opening speech, we hear of the large circle of people involved in the planning and furnishing of the Institute. As regards the scientific equipment, especially Docent H. M. Hansen was ready to help with "thinking through even the smallest of details" but also James Franck's considerable experience was co-opted in the planning phase. We hear also of the architects whose, "inventiveness and patience, have time and again been put to the test", and of the valued contribution from all the craftsmen.

The Institute and all its employees were also, from the beginning, a unified organism with a uniform style. As with other organisms, it has had the need for growth and our task has been to preserve this unity. Moreover, in recent years, we have worked on new organisational forms in order to better exploit the unified initiative of our entire collegial circle. Also here, we have sought to combine theory and experiment in order to learn from human nature.

Today, our thoughts turn particularly to our former colleagues, who over longer or shorter periods, have made valuable contributions to the Institute's work. We are happy to see so many colleagues who now work at other institutes back home again. This relationship has contributed to the strengthening of unity throughout the circle of Danish physicists, which has been a joy to us throughout the years.

The first colleague from another Nordic country was Oscar Klein, who was already involved in the work before the Institute was built and who later, for a long number of years was my father's closest and most invaluable colleague. We are very happy that both Oscar and Gerda Klein could be with us today and we extend a heartfelt greeting to our other Nordic colleagues who have joined us today. The Nordic element of the Institute's life has held a leading position throughout the years and was also the background for the foundation of NORDITA, our sister organisation, with whom we exist in symbiosis and which is of great importance for the atmosphere surrounding the Institute.

Our thoughts are also with the large circle of colleagues, about a thousand in total, from all over the world who have taken part in the Institute's work and who have contributed so great an inspiration. We would have liked to have collected as many participants as possible in order to discuss current problems, but the confines of space at the Institute have made this impossible and such a large gathering would not sit well with our usual forms of discussion. We have therefore had to content ourselves with sending a little greeting to everyone which we also do today.

In the meantime, we can be happy for the thriving connections throughout this circle of colleagues and we have the impression that all feel themselves to be part of a large family. The Institute's close ties to the international community of physicists and the many new talents, who each year come here from all over the world are among the forces to which we attach our greatest hopes for the future.

Father used to say about happiness, that it was a fortunate situation that turned out better that one had in fact deserved, and I think he felt that way about his own work. When we today look back on the last fifty years, we must say that it seems like quite an adventure, one of those where expectations have often been exceeded. On when we look forward, there are plenty of tasks to address and exciting perspectives on the horizon and we have lots of support. Let us hope that in fifty years we will be able to say that fortune favoured us.

CPSIA information can be obtained
at www.ICGtesting.com
Printed in the USA
BVHW070238260819
556602BV00006B/8/P